1991

INTRODUCTION TO
AEROSOL SCIENCE

INTRODUCTION TO AEROSOL SCIENCE

Parker C. Reist

Department of Environmental Sciences and Engineering
School of Public Health
University of North Carolina
Chapel Hill, North Carolina

LIBRARY
College of St. Francis
JOLIET, ILLINOIS

Macmillan Publishing Company
A Division of Macmillan, Inc.
NEW YORK

Collier Macmillan Publishers
LONDON

Macmillan Publishing Company
866 Third Avenue, New York, N.Y. 10022

Collier Macmillan Canada, Inc.

Printed in the United States of America

printing number

1   2   3   4   5   6   7   8   9   10

Library of Congress Cataloging in Publication Data

Reist, Parker C., 1933–
    Introduction to aerosol science.

    Bibliography: p.
    Includes index.
    1. Aerosols.   I. Title.
QC882.R45   1984   541.3'4515   83-42861
ISBN 0-02-949600-4

CONTENTS

PREFACE

"From dust we came and to dust we shall return."

This book is about dust, dust and all the myriad tiny things that hang suspended in the air. These clouds of fine particles, or aerosols, can cheer us up when we look at a spectacular sunset, or can be depressing, such as on a gray day in a smokey town. Particles suspended in air act as sites on which water can condense, and thus play a principal role in the water cycle and the formation of rain. Dust clouds on a back road allow us to follow a vehicle at great distances and smoke screens promise protection in an electronic war. We use fine particles suspended in air to kill mosquitos, treat allergies, control underarm odor, and even oil machinery. High concentrations of some particles are extremely explosive and low concentrations of other particles are extremely toxic. Whether we realize it or not, we are surrounded all the time by literally thousands of small particles, and their importance to the natural functioning of the earth is incalculable.

Considering the importance of airborne particles, one might think they would have continued to attract the attention of scientists and that fundamental knowledge of particle behavior would be widespread by now. This is not the case. Aerosol science is a much neglected stepdaughter of physics or perhaps physical chemistry, and is only now beginning to blossom and provoke the interest it deserves.

Systematic study of the fundamental properties of airborne particles has been intermittent in the past. For some reason we as a society tend to look on everyday phenomena with blind acceptance, regarding what we see as so common that it never occurs to us to ask why. Why does a cloud remain airborne? Where does it come from and where does it go? What is "smoke"? Is it a solid or a gas? When asked on the first day of class many of my students erroneously think smoke is a gas. Why are some dusts harmful and others not? Or similarly, why is the same dust sometimes harmful while other times it is not?

The harmful nature of dust has been known for centuries. At least, early writers indicated in their works a general connection between lung diseases and dust inhalation, even though they didn't distinguish between the various types of respiratory diseases. For example, in the first century A.D. the Roman scholar Pliny referred to inhalation of "fatal dust," and Agricola in the fourteenth century spoke of the "pestilential air" and the "corrosive dust" while writing about mining. In his book published in 1700 Bernardo Ramazzini, an Italian physician, described the effect of dust on the respiratory organs including descriptions of numerous cases of fatal dust disease.

With the industrial revolution in the nineteenth century and the advent of high-speed machinery, dust exposure increased dramatically as did the number of investigators reporting on the effects of the dusty trades. In the latter part of the nineteenth century interest focused on dust exposure of miners, especially in the gold mines of South Africa and the tin mines of Cornwall. As a result of these studies and others it was found that high exposure concentrations gave rise to more cases of lung disease.

Even with evidence showing the relationship of dust levels in the air to disease, there was little effort made to study the properties of dust in the air — how to sample it, how to control it, what its important physical properties are, how it is produced, and where it ultimately goes. It is quite surprising to see the level of misinformation those early nineteenth-century investigators were working with, and it never ceases to amaze me that they were able to deal with the problems as well as they did with the limited knowledge at hand.

Aerosols in the nineteenth century were in the forefront of science because these small particles represented the smallest divisions of matter known. Many individuals whom we now consider to be the intellectual giants of the time contributed to our understanding of aerosols. Tyndall, Lister, Kelvin, Maxwell, Aitken, and Einstein — to mention a few — are familiar names in the aerosol literature as well as in the fields for which they are most famous.

However, with the discovery of radioactivity and the development of quantum mechanics, science, in its quest for the smallest division of the universe, moved away from the study of aerosol particles and the field languished, despite continuing discoveries in medicine regarding the relationship between dust and disease. Only in the area of occupational health were aerosol studies continued, and these were of an applied nature. The use of aerosols in warfare and screening smokes led to some effort to study their properties between the World Wars but it was not until World War II that aerosol problems again began to attract the attention of the main scientific community. The reasons for this increased interest were manifold. First, production of fissionable materials involved working with radioactive aerosols, potentially dangerous materials. Second, the advent of radar raised the need for understanding the effect of clouds on the transmitted and reflected signals and how this effect could be minimized or maximized, depending on whether one wanted to hide or seek. Finally, the threat of chemical and biological warfare needed to be dealt with on the basis of knowledge, not guesswork, and since aerosols represent the chief means for dispensing these agents, study of aerosol behavior was essential.

In the past twenty years work relating directly to the study of aerosols has increased greatly. At least two journals are specifically devoted to aerosol studies and numerous others regularly publish articles on various aspects of particulates in air. Aerosols appear to play a major role in the removal of polluting gases from the atmosphere either by adsorbing them on existing particles or through the creation of new particles. A knowledge of aerosol properties is being found useful in studying the atmosphere of planets other than the earth. Many air pollutants originate in particulate form or become particulates soon after discharge and must be dealt with as such. Acid rain is an example of an aerosol problem where gas is tranformed into a liquid — in this case sulfur dioxide is transformed in the air into sulfuric acid.

As many frustrated investigators have noted again and again, the study of aerosols is by no means easy. Particles in air behave differently than the air in which they are suspended, and behave differently among themselves, depending on their size, shape, and composition. Collecting a representative sample of an aerosol for any purpose can be a frustrating and time-consuming task, and a knowledge of aerosol properties and behavior is essential to maximize chances for adequate sample collection. This is especially true when using many of the automated sampling devices available today. These devices generate the numbers whether they are accurate or not, and it is up to the investigator to interpret and understand what is being generated.

This book is an attempt to present, in a rigorous but illustrative manner, introductory information on the study of aerosol properties and behavior so that an individual who desires to learn the mysteries of the field will not be completely discouraged. The text has evolved out of over fifteen years experience in teaching an introductory course on aerosol science to numerous first year graduate students. Some students picked at the edges of the course and were satisfied; others digested all the material and developed an insatiable appetite for more. I hope in this book to reach both groups. Many example problems have been set to show practical examples of aerosol studies that can be applied almost directly to other situations without paying much attention to the underlying theories. On the other hand, for the more inquiring mind equations have been developed to attempt to illustrate the thought processes used to arrive at particular solutions. Some solutions may not be the most accurate or up-to-date. I have no apologies. In learning the simpler approximate solution one develops the terminology, conventions, and methods of thinking that lead to greater understanding of the more rigorous complex solutions.

This book is a textbook. If it produces individuals who are better able to read and understand the current aerosol literature, and extend the field on their own, it will have served its purpose.

ACKNOWLEDGMENTS

It is difficult for me to begin to list all the people whose effort or encouragement made this book possible. Foremost, perhaps, is Professor Leslie Silverman, who first introduced me to the wonders of aerosols. Then there are my colleagues, at UNC, Donald Fox, Robert Harris, John Hickey, Harvey Jeffries, Arthur Stern, Dan Okun, and David Fraser, and friends elsewhere, Melvin First, Morton Lippman, Mary Amdur, and Sheldon Friedlander, who encouraged me to continue and prodded when the effort slackened. My students have always been supportive, especially Brian Mokler, William Hinds, Matti Jantunen, Mike Kuhlman, Mike Gery, Neil Zimmerman, George Dwiggins, Jim Garrison, and Dave Johnson.

Two of my aerosol classes have seen the text in various draft forms. I appreciate their helpful suggestions and comments, and most especially thank them for pointing out many little errors which somehow seemed to creep in (I hope the reader will find few, if any, of these).

Thanks also to Frances Tindall, who as my editor at Macmillan patiently dealt with missing pages, unanswered comments, and occasional authorly fits of panic.

Several secretaries have had a hand in typing drafts of the text, but it took Donna Simmons to finally organize the effort and keep track of the many details involved in finally getting the text out.

Lastly, without the support of my family I could not have completed this book. Their encouragement is appreciated.

INTRODUCTION TO
AEROSOL SCIENCE

INTRODUCTION AND DEFINITIONS

Aerosols are ubiquitous in our environment. Haze particles are formed over vegetation, dust clouds are blown up by the wind, volcanoes erupt spewing dense smoke into the atmosphere and, of course, man in his many activities marks his way by the particles he discharges into the air. This book is about aerosol particles, their physical properties, and the scientific basis that has been developed for predicting their behavior.

UNITS

Aerosol sizes are usually referred to in terms of the micrometer (μm) (or previously, the micron, μ). One micrometer (micron) is equal to 10^{-4} centimeters, 10^{-6} meters or 10^4 Angstrom units. In working problems it is necessary to use a consistent set of units. Since most physical constants are available either in cgs or mks units (English units are too cumbersome to use), aerosol sizes given in micrometers very often must be converted to either centimeters or meters for computations (depending on the system of units chosen). When working problems involving ratios of particle size this conversion is not necessary.

1

EXAMPLE 1.1

A basketball is 12 in. in diameter. Express its diameter in micrometers.

1 in. = 2.54 cm
1 cm = 10^4 μm

diameter = 12 in. $\times$ 2.54 cm/inch $\times$ 10^4 μm/cm

$\qquad$ = 3.05 $\times$ 10^5 μm

DEFINITIONS

To begin the systematic study of particles, it is first necessary to consider several commonly used definitions of various types of aerosols.

Aerosol A suspension of solid or liquid particles in a gas, usually air; a colloid. Included in this definition would be:

Dust Solids formed by disintegration processes such as crushing, grinding, blasting, and drilling. The particles are small replicas of the parent material, and the particle sizes can range from submicroscopic to microscopic. Very often sizes are specified by screen mesh size. For example, the percentage passing or retained on a given mesh is indicative of size.

EXAMPLE 1.2

How may spherical particles just passing through a 200-mesh screen are required to equal the mass of a single spherical particle that just passes through a 50-mesh screen? Assume that the diameter of the particle passing through the mesh equals the mesh opening and a particle density of 2.65 g/cm^3.

Mass of particle passing 50-mesh screen

$$= [\pi/6] \times d^3 \times \rho = [3.14/6] \times [0.0297]^3 \times 2.65$$
$$= 3.64 \times 10^{-5} \text{ g}$$

Mass of particle passing 200-mesh screen

$$= [\pi/6] \times d^3 \times \rho = [3.14/6] \times [0.0074]^3 \times 2.65$$
$$= 5.62 \times 10^{-7} \text{ g}$$

Number of particles required

$$\frac{3.64 \times 10^{-5}}{5.62 \times 10^{-7}} = 64.7, \text{ say 65 particles}$$

Table 1.1. Openings of some typically small mesh sizes[a]

Mesh	Opening, mm
50	0.297
100	0.150
200	0.074
400	0.038

[a]From Handbook Chem. Phys., 54th Ed., CRC Press, Cleveland, 1973, p.F147.

Fumes Solids produced by physicochemical reactions such as combustion, sublimation, or distillation. Typical fumes are the metallurgical fumes of PbO, Fe_2O_3, or ZnO. Particles making up fumes are quite small, below 1 μm in size, and thus cannot be sized on screens. The particles appear to flocculate readily.

Smoke A cloud of particles produced by some sort of oxidation process such as burning. The optical density is presupposed. Generally, smokes are considered to have an organic origin and typically come from coal, oil, wood, or other carbonaceous fuels. Smoke particles are in the same size range as fume particles.

Mists and fog Aerosols produced by the disintegration of liquid or the condensation of vapor. Because liquid droplets are implied, the particles are spherical. They are small enough to appear to float in moderate air currents. When these droplets coalesce to form larger drops of about 100 μm or so, they can then appear as rain.

Haze Particles with some water vapor incorporated into them or around them, as observed in the atmosphere.

Smog A combination of smoke and fog, usually containing photochemical reaction products combined with water vapor to produce an irritating aerosol. Smog particle sizes are usually quite small, being somewhat less than 1 μm in diameter.

These definitions have arisen from popular usage, so there is little wonder that they overlap. What one person might call smog someone else could call haze, and both would be correct. Therefore we should generally use the more precise, if less colorful, definition of aerosol and then fill in the details on a more quantitative basis.

Since an aerosol is a collection of particles, it is often desirable to indicate whether the particles are all alike or are dissimilar. Thus there are several other descriptors of aerosols that must also be taken into account.

Monodisperse All particles exactly the same size. A *monodisperse aerosol* contains particles of only a single size. As might be expected, this condition is extremely rare in nature.

Polydisperse Containing particles of more than one size.

Homogeneous Chemical similarity. A *homogeneous aerosol* is one in which all particles are chemically identical. In an *inhomogeneous aerosol* different particles have different chemical compositions.

MORPHOLOGICAL PROPERTIES OF AEROSOLS

Shape

It is convenient to think of all aerosol particles as spheres for calculation and this also helps visualize the processes taking place. But, with the exception of liquid droplets, which are always spherical, many shapes are possible. These shapes can be divided into three general classes.

1. *Isometric particles* are those for which all three dimensions are roughly the same. Spherical, regular polyhedral, or particles approximating these shapes belong in this class. Most knowledge regarding aerosol behavior pertains mainly to isometric particles.
2. *Platelets* are particles that have two long dimensions and a small third dimension. Leaves or leaf fragments, scales, and disks fall into this class. Very little is known about platelet behavior in air, and care must be exercised in applying knowledge derived from studying isometric particles to platelets.
3. *Fibers* are particles with great length in one dimension compared to much smaller lengths in the other two. Examples are prisms, needles, and threads or mineral fibers such as asbestos. Recent concern over the health hazard posed by inhalation of asbestos fibers has prompted study of fiber properties in air. There is still not as much known about fibers as isometric particles.

EXAMPLE 1.3

An asbestos fiber is 10 μm in length with a circular cross section of 0.5 μm diameter. Find the diameter of a sphere that has the same volume as the fiber.

$$\text{Volume of fiber} = [\pi/4] \times [0.5]^2 \times 10$$

$$= 1.96 \ \mu\text{m}^3$$

$$\text{Volume of sphere} = [\pi/6] \times d^3$$
$$d^3 = [1.96] \times [6/\pi] = 3.75$$
$$d = 1.55 \, \mu m$$

Particle shape can vary with the formation method and the nature of the parent material. Particles formed by the condensation of vapor molecules are generally spherical in shape, especially if they go through a liquid phase during condensation. Particles formed by breaking or grinding larger particles, termed *attrition,* are seldom spherical, except in the case where liquid droplets are broken up to form smaller liquid droplets.

Size

A particle is generally imagined to be spherical or nearly spherical. Either particle radius or particle diameter can be used to describe particle size. In theoretical discussions of particle properties, the radius is most commonly used, whereas in more practical applications the diameter is the descriptor of choice. Thus one should carefully ascertain which definition is being used when the term particle size is used. In this text particle diameter is used throughout.

Once a choice of diameter or radius is made, there are a number of ways this diameter or radius can be defined which reflect particle properties other than physical size. For a monodisperse aerosol a single measure describes the diameters of all the particles. But with polydisperse aerosols a single diameter is not sufficient to describe all particle diameters, and certain presumptions must be made as to the distribution of sizes. Other parameters besides diameter alone must be used. This is discussed in more detail in Chapter 2.

Two commonly encountered definitions of particle size are Feret's diameter and Martin's diameter. These refer to estimates of approximate particle size when determined from viewing the projected images of a number of irregularly shaped particles. Feret's diameter is the maximum distance from edge to edge of each particle, and Martin's diameter is the length of the line that separates each particle into two equal portions. Since these measures could vary depending on the orientation of the particle, they are only valid if averaged over a number of particles and all measurements are made parallel to one another. Then, by assuming random orientation of the particles, an average diameter is measured.

This measurement problem can be simplified somewhat by using the projected area diameter instead of Feret's or Martin's diameter. This is defined as the area of a circle having the same projected area as the particle in question. Figure 1.1 illustrates these three definitions. In general Feret's diameter will be larger than the projected area diameter which will be larger than Martin's diameter.

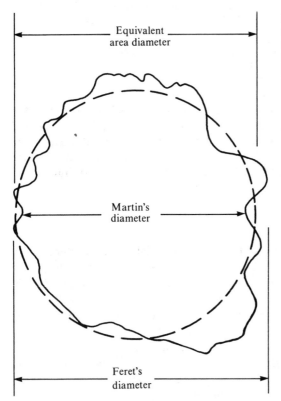

Figure 1.1. Illustration of three common definitions of particle diameter. In general, Martin's diameter is less than the equivalent area diameter, which in turn is less than Feret's diameter.

EXAMPLE 1.4

Figure 1.2 shows a collection of five irregularly shaped particles. By measuring along lines parallel to the scale line, determine Martin's, Feret's, and the projected area diameter for this collection of particles.

The measured values are:

Feret's diameter = 15 scale units;

Martin's diameter = 10 scale units;

Projected area diameter = 13 scale units.

Sometimes a diameter is defined in terms of particle settling velocity. All particles having similar settling velocities are considered to be of the same size, regardless of their actual size, composition, or shape. Two such definitions which are most common are:

Aerodynamic diameter Diameter of a unit density sphere (density = 1 g/cm^3) having the same aerodynamic properties as the particle in question. This means that particles of any shape or density will have the same aerodynamic diameter if their settling velocity is the same.

Stokes diameter Diameter of a sphere of the same density as the particle in question having the same settling velocity as that particle. Stokes diameter and aerodynamic diameter differ only in that Stokes diameter includes the particle density whereas aerodynamic diameter does not.

EXAMPLE 1.5

A sodium chloride cube (density = 2.165 g/cm^3) settles at a rate of 0.3 cm/sec. Find the aerodynamic diameter of this cube.

Appendix A gives a corrected sedimentation velocity of 0.306 cm/sec for a 10-μm-diameter unit density sphere. Hence 10 μm is the aerodynamic diameter of this particular salt cube.

Figure 1.2. Illustration for Example 1.4.

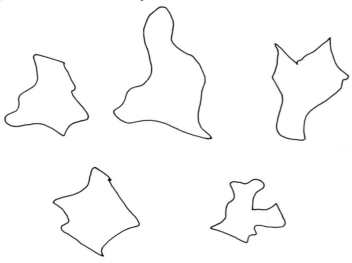

```
0     5    10   15    20
└┴┴┴┴┴┴┴┴┴┴┴┴┴┴┴┘
       Scale units
```

Particle diameters of interest in aerosol science cover a range of about four orders of magnitude, from 0.01 μm as a lower limit to approximately 100 μm as the upper limit. The lower limit approximates roughly the point where the transition from molecule to particle takes place. Particles much greater than about 100 μm or so do not normally remain suspended in the air for a sufficient length of time to be of much interest in aerosol science. There are occasions where particles that are either smaller or larger than these limits are important but usually most particle diameters will fall within the limits of 0.01 μm to 100 μm.

Particles much greater than 5 to 10 μm in diameter are usually removed by the upper respiratory system and those smaller than 5 μm can penetrate deep into the alveolar spaces of the lung. Thus 5 to 10 μm is often considered to be the upper diameter for aerosols of physiological interest.

Within the size range of 0.01 μm to 100 μm lie a number of physical dimensions which have a significant effect on particle properties. For example, the mean free path of an "air" molecule is about 0.07 μm. This means that the air in which a particle is suspended exhibits different properties, depending on particle size. Also, the wavelengths of visible light lie in the narrow band of 0.4 μm to 0.7 μm. Particles smaller than the wavelength of light scatter light in a distinctly different manner than do larger particles.

Particle size is the most important descriptor for predicting aerosol behavior. This is apparent from the above discussion and will become even more apparent in later chapters. Typical particle sizes of selected materials are given below in Table 1.2.

Structure

Aerosol particles may occur by themselves or may be formed into chains of spheres or cubes. These are called *agglomerates* or *flocs*. Agglomerates are usually formed from highly charged small particles such as are found in dense smokes or metal fumes.

Particles may also occur as gas-filled hollow drops or as particle-filled hollow particles. Fly ash is an example of this latter type of material. Thus particle density can be significantly different from the density of the parent material.

Table 1.2. Typical particle diameters (μm)

Tobacco smoke	0.25	Lycopodium	20
Ammonium chloride	0.1	Atmospheric fog	2–50
Sulfuric acid mist	0.3–5	Pollens	15–70
Zinc oxide fume	0.05	"Aerosol" spray products	1–100
Flour dust	15–20	Talc	10
Pigments	1–5	Photochemical aerosols	0.01–1

SURFACE PROPERTIES

Aerosol particles, because of their small size, present a large amount of surface for chemical reactions such as burning, adsorption, absorption, or other chemical reactions, or for such physical properties as wettability or electrostatic effects. The amount of area per gram of material increases as the particle size decreases, and for a given average size, increasing polydispersity decreases the surface area per gram. As particle size becomes very small, the boundary conditions between the particle and the air around it become confused, but also become more important.

EXAMPLE 1.6

What is the surface area of 1 g of a monodisperse water aerosol if the particle diameter is

   a. 10 μm
   b. 1 μm

Let n = particles per gram

$$1 \text{ g} = (\pi/6)(d^3)(1)(n)$$

If A = surface area per gram

$$A = (\pi)(d^2)(n)$$

Then

$$A = \frac{(1)(6/\pi)}{d^3} \times \frac{\pi d^2}{} = \frac{6}{d}$$

For 10-μm particles

$$A = \frac{6}{10^{-3}} = 6000 \text{ cm}^2$$

For 1-μm particles

$A = 60,000 \text{ cm}^2$ $A = 60,000 \text{ cm}^2$

PROBLEMS

1. What is the ratio of the volume of a spherical particle that will just pass through a 200-mesh screen compared to a sphere that will just pass through a 400-mesh screen?

2. Two-tenths of a gram of particles are passed through a 325-mesh sieve but are retained on a 400-mesh sieve. Assuming the particles are spheres and are all of the same size, estimate the maximum and minimum number of particles present. Assume a particle density of 2.65 g/cm^3.

3. Express the earth's equatorial diameter in micrometers ($d = 3963$ miles). Express the diameter of an electron in micrometers ($d = 10^{-12}$ cm). Express the diameter of a hydrogen molecule in micrometers ($d = 2.9$ Å).

4. Compare relative dimensions of a sphere, platelet, and fiber, assuming the fiber element diameter and platelet thickness are one-tenth the sphere diameter and the volumes of the sphere, platelet, and fiber are equal. Assume a circular cross-section.

5. If the sphere in problem 4 is a 1-μm-diameter silica particle ($\rho = 2.65$ g/cm^3) what would be the equivalent platelet dimensions and fiber length?

6. The settling velocity of a 5-μm-diameter sand particle can be estimated from the expression

$$v_g = 3 \times 10^{-3} \, d^2 \, \rho$$

where v_g is the settling velocity in cm/sec, d the particle diameter in μm, and ρ the density in g/cm^3.

Find the aerodynamic diameter of this particle.

7. Show that for a constant mass of particles, decreasing the particle size by a factor of 10 increases the surface area by a factor of 10.

8. How many 0.1-μm-diameter H_2SO_4 droplets can be produced by splitting up up one 10-μm-diameter H_2SO_4 droplet?

PARTICLE SIZE DISTRIBUTIONS

INTRODUCTION

As mentioned in Chapter 1, most frequently aerosol particles are present in a variety of sizes, that is, the aerosol is *polydisperse.*

Most aerosols are polydisperse when formed, some more than others. For example, an examination of sawdust would reveal particles of various sizes, as would any material formed by attrition. Since raindrops could grow by condensation or by a series of collisions with other drops, they would also be expected to be polydisperse. In fact, monodisperse aerosols are very rare in nature and when they do appear, generally do not last very long. Some high-altitude clouds are monodisperse, as are some materials formed by condensation. Sometimes it is satisfactory to represent all the particle sizes only by a single size. Other times more information is needed about the distribution of all particle sizes. Of course, a simple plot of particle frequency versus size gives a picture of the sizes present in the aerosol but this may not be enough for a complete quantitative analysis.

There are a number of ways polydisperse aerosols can be described, using mathematical or visual methods. Some of the more common methods are discussed this chapter.

MEAN AND MEDIAN DIAMETER

The simplest way of treating a group of different particle diameters is to add up all the diameters and divide by the total number of particles. This gives the average diameter. Mathematically this can be expressed as:

$$\bar{d} = \frac{\Sigma\, n_i\, d_i}{\Sigma\, n_i} \qquad (2.1)$$

This is known as the *mean particle diameter.*

The *median particle diameter* can be determined by listing all diameters in order from the smallest to the largest and then finding the particle diameter that splits the list into two equal halves.

EXAMPLE 2.1

Given the following particle diameter data, determine the mean and median diameters of the aerosol.

Interval, μm	Number, n_i
1 - 2	30
2 - 3	90
3 - 5	50
5 - 10	20
10 - 20	10

Using Eq. (2.1) the following table can be formed. The midpoint of the size interval is chosen as the best estimate of the size of all particles in that interval.

Midpoint, $= d_i$	n_i	$n_i\, d_i$
1.5	30	45
2.5	90	225
4.0	50	200
7.5	20	150
15.0	10	150
	200	770

The mean value is [770/200] = 3.85 μm

By inspection of the table, the median value can be seen to lie somewhere between 2 and 3 μm in diameter. With the data available a more precise evaluation of this number is not possible.

Although they are simple in concept, neither the mean nor median diameter alone conveys much information about the general range of particle diameters present. Usually more information is required describing the spread of the particle size distribution. This gives some indication about how well the mean or median value represents all particles in the aerosol.

It is a common practice to describe an aerosol solely by some average value, completely ignoring considerations of particle size distribution. When this is done estimates of aerosol properties are much less accurate than they would be if all particle sizes were taken into account.

HISTOGRAMS

Besides determining a mean or median value, the number of particles in various size intervals can be plotted as bar charts or line charts. These plots are pictures of the size distribution of the aerosol. This is useful in envisioning the range and frequency of the sizes present.

EXAMPLE 2.2

Plot the data given in Example 2.1 first by plotting the midpoints of the size intervals as a function of particle diameter and then by plotting a bar chart of number of particles per unit size interval against each size interval.

Interval, μm	Midpoint, $= d_i$	Interval size, μm	n_i	$n_i/\mu m$
1 – 2	1.5	1	30	30
2 – 3	2.5	1	90	90
3 – 5	4.0	2	50	25
5 – 10	7.5	5	20	4
10 – 20	15.0	10	10	1

Figure 2.1(a) shows a line chart plot of the midpoints of the data. Although the particle diameter distribution is plainly shown, it is possible to alter the shape of the distribution by changing the interval size.

When a bar chart is plotted instead of a line chart, Fig. 2.1(b), this problem is not as severe. The ordinate or height of each bar is normalized by dividing the number of particles in an interval by the width of that interval. The width of each bar represents the actual width of each size interval. Then the area of each block represents the relative frequency of particles in that particular size interval.

Charts or graphs of this sort have the advantage of showing at a glance what the particle size distribution of an aerosol looks like and are perhaps the best way of visually representing complex size-distribution data.

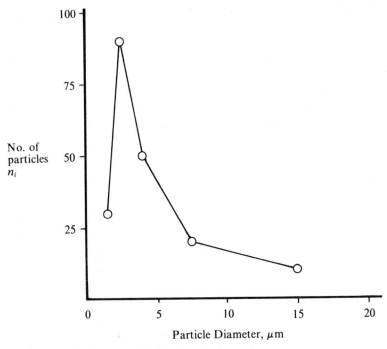

Figure 2.1a. Simple plot of distribution data.

For atmospheric aerosols a wide range of particle sizes may be present in numbers which can vary by several orders of magnitude. In these cases the typical bar graph will not be satisfactory since the large numbers of small particles can completely overwhelm the display of other sizes, even though the larger sizes may be most significant in terms of mass or surface area. Or the larger particle diameters will be displayed more prominently than the smaller ones, even though the smaller sizes may be of primary interest.

One solution is to plot the logs of particle diameter on the abcissa instead of the diameters themselves. This spreads out the presentation of distribution data so that a much broader range of particle sizes can be visualized. However, to maintain the relationship that the area between two particle size intervals is proportional to the total number of particles present, the ordinate scale must be altered. This is done by dividing the number of particles in each interval by the difference in the logarithms of the largest and smallest particle sizes of that interval, or, in mathematical terms:

$$\text{Ordinate value} = \frac{\Delta n}{\Delta \text{Log } d} \tag{2.2}$$

This relationship is found for each size interval. Similar expressions can be written for particle surface area or particle mass or volume. (It should be stressed

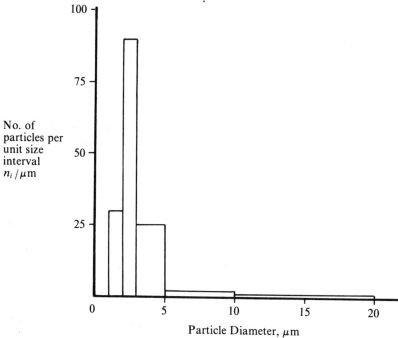

Figure 2.1b. Bar graph of particle distribution data.

again that particle volume converts directly to particle mass by multiplication of volume times particle density. Hence in plotting size distribution data, either one can be used to represent the other.)

EXAMPLE 2.3

Plot the data given in Example 2.1 in the form of $[\Delta n/\Delta \log d]$ versus $\log d$.

Interval	Midpoint	No., Δn	$\Delta \log d$	$\Delta n/\Delta \log d$
1 – 2	1.5	30	0.30	100
2 – 3	2.5	90	0.18	500
3 – 5	4.0	50	0.22	227
5 – 10	7.5	20	0.30	67
10 – 20	15.0	10	0.30	33

The data are plotted on Fig. 2.2. A continuous distribution is assumed in order to develop a smooth plot.

Continuous curves of the type illustrated in Fig. 2.2 are often used to show the difference in size distributions of aerosol number, surface area, or mass, with

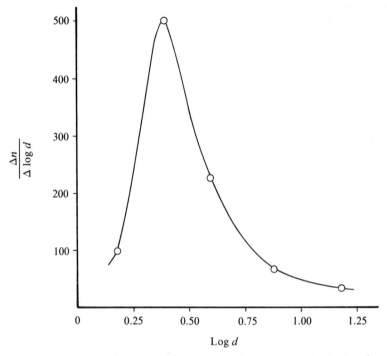

Figure 2.2. Plot of data from Example 2.3. A continuous distribution is assumed.

the same aerosol. These differences arise when there are large numbers of small particles present in an aerosol. These particles contribute greatly to total particle count but little to total particle mass or surface area.

EXAMPLE 2.4

Plot surface and volume distribution for the aerosol given in Example 2.1 in the same fashion as plotted in Example 2.3. However in this case normalize the linear ordinate so that the relative areas and volumes under similar interval limits will be comparable.

For particle surface area (S), values in the ΔS column are determined by multiplying the number of particles each interval by the square of the midpoint diameter of that interval. For particle volume (V), values in the ΔV column are found by multiplying the number of particles in each interval by the cube of the midpoint particle diameter. It is not necessary to multiply the ΔS values by π or the ΔV values by $\pi/6$ since these constants will cancel out when the ΔS and ΔV quantities are normalized by dividing by the sum of all values.

Interval	ΔS	$\Delta S/S_T \,\Delta \log d$	ΔV	$\Delta V/V_T \,\Delta \log d$
1 – 2	67.5	0.047	101.3	0.007
2 – 3	562.5	0.650	1406.3	0.167
3 – 5	800.0	0.757	3200.0	0.310
5 – 10	1125.0	0.780	8437.5	0.600
10 – 20	2250.0	1.561	33750.0	2.399

$$S_T = 4805.0 \qquad\qquad V_T = 46{,}895.1$$

Comparison of the data plotted in Fig. 2.3 shows how surface area and volume (mass) tend to be associated mainly with the larger size particles whereas in general the smaller particles contribute mainly to the total numbers present. Therefore in presenting size distribution data it is important to consider the purpose of the presentation and which feature (number, surface area or volume [mass]) is to be stressed.

Figure 2.3. Similar plot as in Figure 2.2 except particle surface area, S, and particle volume, V, are plotted. These curves are plotted. These curves are normalized by dividing by S_T and V_T, respectively.

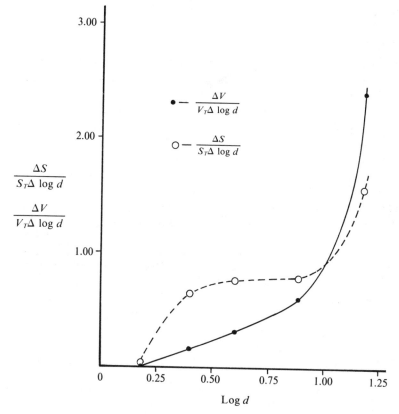

MATHEMATICAL REPRESENTATION OF DISTRIBUTION

If the size interval of the aerosol is permitted to become very small, the resulting histogram begins to approximate a smooth curve. Then it is possible to represent the distribution by a smooth curve or, better still, by some mathematical function, i.e.,

$$dn_i = f(d)\, dd \tag{2.3}$$

where dn_i is the number of particles lying in the interval between sizes dd_{i-1} and dd_i.

Obviously, to plot this sort of curve requires analysis of the sizes of a great number of particles. Or, if it were possible to specify some identifying parameters of the distribution, a functional form could be used to represent a whole family of curves. There have been many attempts to find such a functional form. Usually these equations have been satisfactory for aerosols from the same specific sources but are not generally applicable to all aerosols.

One widely used form which is applicable to many different aerosols from a variety of sources is the *log-normal distribution*. To understand the utility of the log-normal distribution it is first necessary to review the concept of a "normal" distribution.

Normal Distribution

Many phenomena which appear to happen on a more or less random basis exhibit certain characteristics which can be used to predict future trends. For example, although it is impossible to tell on any single toss whether a coin will come up heads or tails, if the coin is unbiased heads will come up approximately 50% of the time. The more tosses made, the closer one usually comes to this approximation.

Suppose 100 tosses were made and the number of heads recorded, and the experiment repeated many times. It would be observed that although usually there would be about 50 heads for every 100 tosses, occasionally there would be substantially more or less. If the frequency of heads were plotted as a function of the number of heads observed in 100 tosses, a curve shape would be found that is entirely predictable. This shape, known as a "normal" distribution or "normal" curve, is shown in Fig. 2.4a. The primary virtue of a normal distribution is that because it is predictable, it can be described with two characteristic numbers, a mean value and a standard deviation. These are shown in Fig. 2.4a and are defined mathematically as:

$$\bar{d} = \frac{\displaystyle\sum_{i=0}^{\infty} n_i d_i}{\displaystyle\sum_{i=0}^{\infty} n_i} \tag{2.4}$$

$$\sigma = \left[\frac{\sum\limits_{i=0}^{\infty} n_i(\bar{d} - d_i)^2}{\sum\limits_{i=0}^{\infty} n_i - 1} \right]^{1/2}$$

(2.5)

EXAMPLE 2.5

Compute the value of the standard deviation, σ, for the data given in Example 2.1.

From Example 2.1 the mean value was determined to be 3.85 μm.

Interval	Midpoint	No., n_i	$(\bar{d} - d_i)$	$(\bar{d} - d_i)^2 \times n_i$
1 - 2	1.5	30	2.35	165.68
2 - 3	2.5	90	1.35	164.03
3 - 5	4.0	50	-0.15	1.13
5 - 10	7.5	20	-3.65	266.45
10 - 20	15.0	10	-11.15	1243.23
		200		1840.52

The standard deviation, $\sigma = [1840.52/(200 - 1)]^{1/2} = [9.25]^{1/2} = 3.04$ μm.

Figure 2.4a. The "normal" distribution.

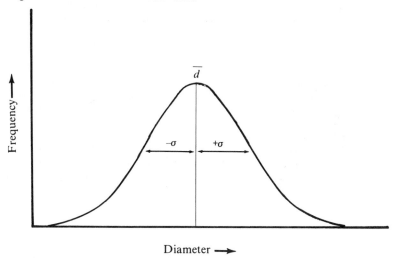

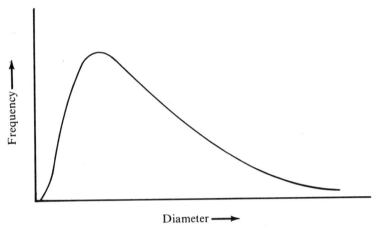

Figure 2.4b. The "log-normal" distribution.

Means and standard deviations can be calculated for any set of data. For data which are normally distributed, however, the mean value lies at the midpoint of the data (hence it is also the median), and 67% of the distribution falls between the range of plus or minus one standard deviation.

Normal distributions occur with a variety of statistical data including average height and weight of children, grade distributions of large groups of students, and even frequency of underweight or overweight candy bars on a production line. One might guess that aerosol particle sizes would also be normally distributed.

Unfortunately, this is generally not the case. For many aerosols a plot of frequency versus size results in a graph similar to that shown in Fig. 2.4b, in which there are proportionally many more smaller particles than larger ones. The curve is said to be skewed toward the larger particle sizes.

Log-normal Distribution

It was observed many years ago that particle size data which were skewed and did not fit a normal distribution would very often fit a normal distribution if frequency were plotted against the *log* of particle size instead of particle size alone. This tended to spread out the smaller size ranges and compress the larger ones. If the new plot then looked like a normal distribution the particles were said to be *log-normally* distributed and the distribution was called a *log-normal* distribution. By analogy with a normal distribution the mean and standard deviation became:

$$\log d_g = \frac{\Sigma n_i \log d_i}{\Sigma n_i} \qquad (2.6)$$

known as the *geometric mean diameter,* and

$$\log \sigma_g = \left[\frac{\Sigma n_i (\log d_g - \log d_i)^2}{\Sigma(n_i) - 1} \right]^{1/2} \tag{2.7}$$

where σ_g is known as the *geometric standard deviation.*

EXAMPLE 2.6

Compute the geometric mean diameter (d_g) and geometric standard deviation (σ_g) for the data given in Example 2.1.

To compute the geometric mean

Interval	d_i	No., n_i	log d_i	$n_i \times$ log d_i
1 - 2	1.5	30	0.176	5.283
2 - 3	2.5	90	0.398	35.815
3 - 5	4.0	50	0.602	30.103
5 - 10	7.5	20	0.875	17.501
10 - 20	15.0	10	1.176	11.761
		200		100.463

The geometric mean, $d_g = \log^{-1} [100.463/200] = \log^{-1} 0.502 = 3.18 \, \mu m$

To compute geometric standard deviation

d_i	No., n_i	log d_g - log d_i	$n_i \times$ (log d_g - log d_i)^2
1.5	30	0.326	3.195
2.5	90	0.104	0.983
4.0	50	-0.100	0.496
7.5	20	-0.373	2.777
15.0	10	-0.674	4.538
	200		11.989

The geometric standard deviation, $\sigma_g = \log^{-1} [11.989/199]^{0.5}$

$$= \log^{-1} 0.245 = 1.760$$

Notice that σ_g is a pure number. Unlike the regular standard deviation it has no units. This is because it represents a ratio of diameters.

With a log-normal distribution, one geometric standard deviation represents a range of particle sizes within which lie 67% of all sizes. In this case the range is from $[d_g/\sigma_g]$ to $[d_g \times \sigma_g]$, unlike the simple additive case for a normal distribution. Ninety-five percent of all particles would lie in a range $[d_g/2\sigma_g]$ to

$[d_g \times 2\sigma_g]$. Thus for a monodisperse aerosol, σ_g is equal to 1 whereas σ is equal to zero for a normal distribution.

The functional form of the log normal distribution can be written as (Herdan, 1960)

$$f(d) = \frac{1}{d \ln \sigma_g (2\pi)^{0.5}} \exp \left[-\frac{(\ln d - \ln d_g)^2}{2 \ln^2 \sigma_g} \right] \tag{2.8}$$

where

$$\int_0^\infty f(d)\, dd = 1 \tag{2.9}$$

EXAMPLE 2.7

Given $d_g = 1\ \mu m$ and $\sigma_g = 2$. Find $f(d)$ when $d = d_g$.

Solution:

$$f(d) = \frac{1}{d \ln \sigma_g (2\pi)^{0.5}} \exp \left[-\frac{(\ln d - \ln d_g)^2}{2 \ln^2 \sigma} \right]$$

$$= \frac{1}{(1) \ln 2\ (2\pi)^{0.5}} \exp \left[-\frac{(\ln 1 - \ln 1)^2}{2(2)^2} \right]$$

$f(d) = 0.576\ \mu m^{-1}$

The approximate fraction of particles lying within the range $0.95\ \mu m$ to $1.05\ \mu m$ would be $0.576 \times 0.1 = 0.058 = 5.8\%$

LOG-PROBABILITY PAPER

Because a log-normal distribution can be expressed as a distinct mathematical function, it is possible to construct graph paper on which a cumulative log-normal distribution plots as a straight line. An example of such a plot is shown in Fig. 2.5. Data are plotted as cumulative percentage of particles equal to or less than the largest size of each size interval versus the upper size of that size interval. A straight line on such a plot implies a log-normal distribution.

If a straight line can be fitted to the plot, then the median particle diameter can be determined as being the 50% value on the plot (remember that when plotting number distribution, geometric mean and median for the number distribution are the same if there is a log normal distribution). The geometric standard deviation is determined by the ratio:

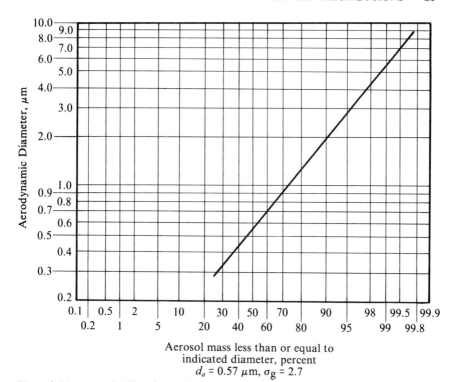

Figure 2.5.   Log probability plot.

$$\sigma_g = \frac{84.13\% \text{ diameter}}{50\% \text{ diameter}} = \frac{50\% \text{ diameter}}{15.87\% \text{ diameter}} \qquad (2.10)$$

Use of log probability paper is the simplest way to determine the mean and geometric standard deviation provided *the distribution does indeed follow a log-normal shape, or at least approximates it.*

OTHER DEFINITIONS OF MEANS

There are a number of different mean values or median values which can be defined for a particle size distribution. For example, besides the common count mean diameter, diameter of average mass can be defined as representing the diameter of a particle whose mass times the number of particles per unit volume is equal to the total mass per unit volume of the aerosol. Or diameter of average surface can be defined in a similar way for the particle whose surface area times the number of particles per unit volume equals the total surface area per unit volume in an aerosol. For a log-normally distributed aerosol different diameters

Table 2.1. Value of p for various diameter definitions

Definition Number	To Get	Let p Equal
1	Mode	−1
2	Geometric mean or median	0
3	Arithmetic mean	0.5
4	Diameter of average area[a]	1
5	Diameter of average mass	1.5
6	Surface median diameter	2
7	Surface mean diameter[b]	2.5
8	Volume median diameter	3
9	Volume mean diameter	3.5

[a]Defined as $\left[\dfrac{\Sigma n_i\, d_i^2}{\Sigma n_i}\right]^{0.5}$

[b]Defined as $\log^{-1}\left[\dfrac{\Sigma n_i\, (\log d_i)^2}{\Sigma n_i}\right]^{0.5}$

can be related by the equation (Raabe, 1971):

$$d_p = d_g \exp (p \ln^2 \sigma_g) \tag{2.11}$$

where p is a parameter which defines the various possible diameters.

Table 2.1 gives the relationship of the parameter p to the various definitions of aerosol particle diameter. Figure 2.6 shows the relative location of each of these diameters on a typical log-normal distribution plot.

This relationship is a more general form of a well-known relationship used for converting particle number measurements to mass measurements and vice-versa known as the Hatch-Choate equation (Drinker and Hatch, 1954).

Figure 2.6. Example of log-normal distribution. Definition of terms: $d_g = 1.00$, $\Sigma_g = 2.00$.

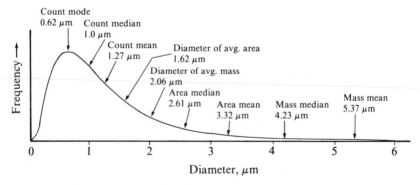

In its original form the Hatch-Choate equation for conversion of number to mass is given by:

$$\log d_{mmd} = \log d_g + 6.9 \log^2 \sigma_g \qquad (2.12)$$

where d_{mmd} is the mass median diameter (No. 8 of Table 2.1), and for surface median diameter, d_{smd} (No. 6 in Table 2.1),

$$\log d_{smd} = \log d_g + 4.6 \log^2 \sigma_g \qquad (2.13)$$

These equations can be derived from Eq. 2.11. It is important to note that σ_g will be the same regardless of the definition of diameter used. That is, with a log-normal distribution σ_g will be the same whether number, surface, or mass mean diameters are being measured:

EXAMPLE 2.7

Given a log-normally distributed aerosol with a geometric mean diameter of 1.5 μm and a σ_g of 2.3, what is the surface area median diameter and the mass median diameter of this aerosol?

Using the Hatch-Choate equation for surface median diameter

$$\log d_{smd} = \log d_g + 4.6 \log^2 \sigma_g$$
$$\log d_{smd} = 0.176 + [4.6] \times [0.362]^2$$
$$\log d_{smd} = 0.176 + 0.602 = 0.778$$
$$d_{smd} = 6.0 \ \mu\text{m}$$

Mass median diameter will be the same as volume median diameter (since particle density cancels out in computing the means. Thus we can use the Hatch-Choate relationship directly:

$$\log d_{mmd} = \log d_g + 6.9 \log^2 \sigma_g$$
$$\log d_{mmd} = 0.176 + [6.9] \times [0.362]^2$$
$$\log d_{mmd} = 1.080$$
$$d_{mmd} = 12.0 \ \mu\text{m}$$

With a log-normal distribution the volume or mass median diameter will always be greater than the surface median diameter which will in turn be greater than the number median diameter.

PROBLEMS

1. Given the following data:

Size interval, μm	Number
0.1 – 0.5	120
0.5 – 0.8	380
0.8 – 1.4	146
1.4 – 2.7	96
2.7 – 5.6	53
5.6 – 8.9	22
8.9 – 12.6	8

construct a histogram showing number per unit size interval for each size interval. Show that the area of each block is proportional to the number of particles represented by that block.

2. Using the data in Problem 1, compute the mean particle diameter and standard deviation of this distribution.

3. Using the same data, compute a geometric median size and geometric standard deviation. What would be the numerical value of the geometric standard deviation if the particles were all of the same size?

4. With the data given in Problem 1, plot the number distribution function, $[\Delta n/(n_T \, \Delta \log d)]$, and the mass distribution function, $[\Delta m/(m_T \, \Delta \log d)]$, as a function of log of particle diameter. Assume all particles within a size interval are spheres having a diameter equal to the mid-point of the size interval. The density of the particles equals 1 g/cm^3.

5. Plot the data from Problem 1 on log probability paper. Find the line of best fit of these data and then determine the geometric mean and geometric standard deviation from this line.

6. Using the Hatch-Choate equation, compute the mass median diameter from the information developed in Problem 5. If the aerosol contains one million particles per cubic foot and the particle density is 1 g/cm^3, find the aerosol concentration in $\mu g/m^3$.

7. Show that the Hatch-Choate equations are just special cases of the general equation for log-normal distributions, Eq. 2.11.

8. Show that the integral of $f(x)$ for a log-normal distribution (Eq. 2.8) does equal 1 when $f(x)$ is integrated over the limits of zero to infinity.

THREE

FLUID PROPERTIES

An aerosol is a suspension of particles in a gaseous medium. Without the medium there would be no aerosol. The medium acts to restrain random particle motion, supports the particles against the strong pull of gravity, and in some cases acts as a buffer between particles. It is impossible to properly study aerosol behavior without first considering the medium in which the particles are suspended.

Medium behavior can be visualized in two ways. First, it can be considered to be a large collection of small spheres (molecules) that are in random motion with each other but may be in ordered motion overall. A general treatment of matter from a molecular point of view is called *statistical mechanics,* the nonequilibrium gaseous portion is referred to as *kinetic theory.*

A second way to visualize gas behavior is by considering the gas to be a *continuous medium,* that is, similar to some sort of interlocking syrup such as molasses or water. Study of medium properties in this case is known as *fluid dynamics,* or for air, *aerodynamics.* In the first case the microscopic (small) properties of the gas are important. In the second, it is the macroscopic (large) properties which are of interest. Since aerosol particles can span the range from near-molecular sizes up to hundreds of micrometers, the gas in which the particles are suspended must be considered both from a molecular point of view and as a continuous medium.

In studying aerosols it is important to develop in one's mind's eye a picture of the process taking place. By visualizing the problem (even if it is in a simplified form) it is easier to find a method of solution, since most problems are more difficult to set up than they are to solve once stated. To carry out this visualization one must have an understanding of the physical phenomena that come into play and a means for estimating their effect. Thus when considering a pitched baseball it is only necessary to visualize how hard the ball is thrown, and whether any spin is imparted to it. If this baseball were 1 μm in diameter, it would also be extremely important to consider the properties of the air through which the baseball travels. Why? Because the medium looks different to the baseball-sized baseball than to the micrometer-sized baseball. In this chapter various properties of the medium (usually air, but it could be any other gas) will be discussed from both a microscopic and macroscopic viewpoint, so that visualization skills can be enhanced and important medium properties introduced.

KINETIC THEORY

The following represents only the briefest discussion of kinetic gas theory. For more information there are many good texts on the subject (e.g., see Daniels and Alberty, 1961).

In considering a gas from the molecular point of view three main assumptions can be made initially (Daniels and Alberty, 1961). These are:

1. The gas volume of interest contains a very large number of molecules.
2. The molecules are small compared to the distances between them and are in a state of continuous motion traveling in straight lines between collisions.
3. The molecules are spherical and do not interact with each other except by elastic collisions. Elastic collisions represent no energy loss due to rearrangement of the interior of the molecule.

With these assumptions it is possible to simplify molecular behavior to a point where the gas can be treated statistically.

EXAMPLE 3.1

Determine the number of molecules in 1 cm^3 of air at 760 mm pressure, 20°C.

Let V = volume of gas occupied by one mole = 22.4 liters at standard conditions.

For 20°C this volume must be increased in proportion to the increase in absolute temperature.

Zero degrees centigrade on the absolute scale is 273° Kelvin. Thus

$$V_{20°C} = 22.4 \times \left[\frac{273° + 20°}{273°}\right] = 24.04 \text{ liters}$$

Then, since the number of molecules in one mole is 6.02×10^{23}, i.e., *Avagadro's number,*

$$\frac{A}{V} = \frac{6.02 \times 10^{23}}{24.04 \times 10^3} = 2.50 \times 10^{19} \text{ molecules/cm}^3$$

Example 3.1 illustrates the large number of molecules that are present even in a fairly small volume of gas. Thus the first assumption holds. In dealing with very low pressures (small numbers of molecules per unit volume) or small volumes, statistical assumptions may not hold and in these cases it is important to consider the medium properties quite carefully before applying any generalities about aerosol behavior.

EXAMPLE 3.2

Assuming all molecules are regularly spaced within a 1-cm^3 volume, determine the average distance between them.

If there are 2.50×10^{19} molecules/cm^3, then there would be one molecule for each

$$\frac{1}{2.50 \times 10^{19}} \text{ cm}^3$$

$$= 4.0 \times 10^{-20} \text{ cm}^3$$

This represents a cube surrounding a single molecule. The length of one side of the cube or the distance between two molecules is

$$\text{distance} = (4.0 \times 10^{-20})^{1/3}$$

$$= 3.42 \times 10^{-7} \text{ cm}$$

$$= 34.2 \text{ Angstrom units, Å}$$

Typical molecular diameters for gas molecules range from about 2 to 5 Å. Hence it can be concluded that the second assumption holds, since even with this simplistic analysis the average distance between molecules is at least ten times the molecular diameters.

For aerosols, the smallest particle diameters are about 0.005 μm increasing to 100 μm or so (50 Å to 1,000,000 Å). At the smallest sizes, aerosol particles begin to approach some very large molecules in size.

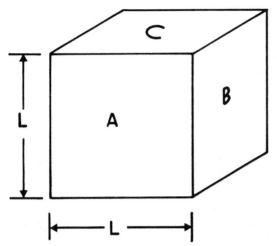

Figure 3.1. Box envisioned for molecular speed estimates.

GAS BEHAVIOR

Some of the basic properties of gases can be deduced using fairly simple logic. Since this same sort of reasoning is used later to deduce aerosol properties, it is instructive here to give two examples of the types of thought processes that yield great insight into physical phenomena.

Molecular Speeds (Bernoulli)

This is a very simple approach to the question of how fast the molecules are moving in a gas.

Let N molecules be enclosed in a cubical box as illustrated in Fig. 3.1, the length of each edge being L. We assume that one third of the molecules move back and forth so that they strike Face A, one third move in a similar manner so that they strike Face B, and one third move similarly so that they strike Face C.

After a molecule strikes one of the faces, say Face B, it must travel a distance of $2L$ before it strikes Face B again. Therefore it makes $C/2L$ hits per unit time, where C is the velocity of the molecule. If the mass of the molecule is m, at each hit the molecule imparts a momentum of $2mC$. (The factor 2 comes from the molecular velocity changing from $+C$ to $-C$.)

The change in momentum per unit time, dp/dt, is the change per hit times the number of hits;

$$\frac{dp}{dt} = (2mC) \times \left(\frac{C}{2L}\right) = \frac{mC^2}{L} \tag{3.1}$$

The total momentum transferred to Face B in unit time is

$$\left(\frac{N}{3}\right)\frac{mC^2}{L} \qquad (3.2)$$

Newton's Second Law of Motion states that force is proportional to the rate of change of momentum. Therefore the total momentum transferred to the wall per unit time is equal to the force acting on that wall.

$$F = \left[\frac{1}{3}\right]\frac{NmC^2}{L} \qquad (3.3)$$

$$\text{Pressure} = P = \frac{\text{force}}{\text{area}} = \frac{NmC^2}{3L^3} = \frac{NmC^2}{3V} \qquad (3.4a)$$

where V is the volume of the box.

Since (Nm/V) is the density, ρ, of the gas,

$$P = \left[\frac{1}{3}\right]\rho C^2 \qquad (3.4b)$$

or

$$C = (3P/\rho)^{1/2} \qquad (3.5)$$

Thus it is possible to compute the average speed of gas molecules merely from a knowledge of the pressure, P, and gas density, ρ. For hydrogen under standard conditions, $C = 1696$ m/sec, approximately the speed of a bullet. This simple derivation is reasonably accurate, even though the assumption is made that all molecules are traveling at the same velocity. Often simplifying assumptions permits the parameters in an equation to be identified, even if the values of the constants may be somewhat inaccurate.

EXAMPLE 3.3

Compute the estimated speed of an "air" molecule at 20°C, normal pressure.

$$\text{Density air} = \frac{MW}{MV} = \frac{29}{22.4 \times 10^3} \times \frac{(273)}{(293)}$$

$$= 1.21 \times 10^{-3} \text{ g/cm}^3$$

Atmospheric pressure = 760 mm Hg

$$= 1013.25 \text{ mb}$$

$$C = (3[1013.25 \times 10^3]/1.21 \times 10^{-3})^{1/2}$$

$$= (2.51 \times 10^9)^{1/2}$$

$$= 501 \times 10^2 \text{ cm/sec}$$

The derivation presented above gives a reasonably close estimate of actual molecular velocities, despite its obvious simplifications.

In actuality, molecular velocities are not all the same. At any time some molecules are moving much faster than the average while others are moving more slowly than the average. For a perfect gas the velocity distribution (in one dimension) is given by the Maxwell-Boltzmann distribution function,

$$f(v_x)\, dv_x = \left(\frac{m}{2\pi kT}\right)^{1/2} \times \exp\left(-mv_x^2/2kT\right) dv_x \qquad (3.6)$$

A plot of this equation is shown in Figure 3.2a. The term k is Boltzmann's constant, $k = 1.38 \times 10^{-16}$ erg/° Kelvin.

The most probable velocity in the x-direction is zero with positive and negative velocities having equal probabilities.

EXAMPLE 3.4

Find the most probable value of $f(v_x)$ for an air molecule. Assume normal temperature and pressure.

The most probable value of $f(v_x)$ occurs when $v_x = 0$. Hence

$$f(v_x) = \left[\frac{m}{2\pi kT}\right]^{1/2} = \left[\frac{\dfrac{29}{6.02 \times 10^{23}}}{(2\pi)\,1.3e8 \times 10^{-16}(293)}\right]^{1/2}$$

$$f(v_x) = 1.38 \times 10^{-5}$$

Figure 3.2a. One-dimensional Maxwell-Boltzmann velocity distribution.

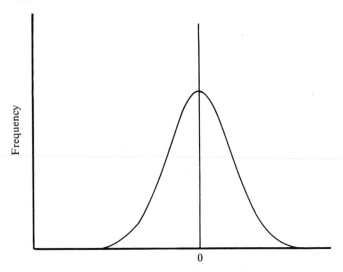

Velocity in One Direction, V_x

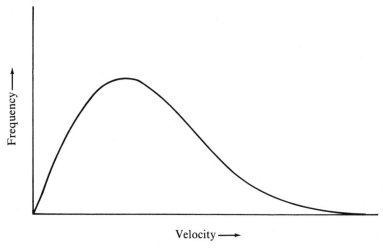

Figure 3.2b. Three-dimensional Maxwell-Boltzmann velocity distribution.

Although Eq. 3.6 represents molecular behavior in a single direction, when all three directions are taken into account simultaneously the probability that an arbitrarily selected molecule will have a velocity between v and $v + dv$ is

$$f(v)\, dv = \left[\left(\frac{m}{2\pi kT}\right)^{3/2} \times \exp\left(-mv^2/2kT\right)\right] \times \left[4\pi v^2\right]\, dv \qquad (3.7)$$

Equation (3.7) says that the probability of having zero velocity is zero. That is, there is no chance that at any time a molecule will completely stop in its motion. Fig. 3.2b shows a plot of Eq. (3.7) for air. There are several ways in which the representative velocity of the ensemble of molecules can be defined.

The arithmetic mean velocity is

$$\bar{v} = \int_0^\infty v\, f(v)\, dv = \left[\frac{8kT}{\pi m}\right]^{0.5} \qquad (3.8)$$

The most probable velocity is obtained by taking the derivative of $f(v)\, dv$ with respect to v and setting it equal to zero.

$$v_p = \left[\frac{2kT}{m}\right]^{0.5} \qquad (3.9)$$

The root-mean-square velocity is

$$(\overline{v^2})^{0.5} = \left[\int_0^\infty f(v)\, v^2\, dv\right]^{0.5} = \left[\frac{3kT}{m}\right]^{0.5} \qquad (3.10)$$

Each of these average values can be used to best describe some particular property of the ensemble of molecular velocities. For example, in a gas all molecules have the same average kinetic energy. Hence, the root-mean-square velocity is the best estimate of velocity to use for computing parameters that are a function of kinetic energy,

$$\bar{E} = \frac{m\overline{v^2}}{2} = \frac{m3kT}{2m} = \frac{3kT}{2}$$

(3.11)

The term $\bar{E}$ is the average energy of a molecule in a gas. Interestingly, an aerosol particle suspended in the gas will acquire this same average kinetic energy from the molecules in the gas.

EXAMPLE 3.5

What is the average kinetic energy of a 1.0-μm unit density sphere which is in equilibrium with its surroundings? The air temperature is 20°C.

$$E = \frac{3kT}{2} = \frac{(3)(1.38 \times 10^{-16})(293)}{2}$$

$$E = 6.07 \times 10^{-14} \text{ ergs}$$

Mean Free Path

The mean free path is defined as the average distance a molecule will travel in a gas before it collides with another molecule. This is related to molecular spacing but takes into account the fact that all molecules are in a constant state of motion and thus are more widely separated than they would be if they were firmly bound to each other. Mean free path can be estimated using the following simple argument.

Consider a molecule traversing the center line of a tunnel whose diameter, 2σ, is equal to twice the molecule diameter, σ (Fig. 3.3). The molecule will collide with all molecules whose centers lie a distance of σ away from the centerline of the tunnel and miss all others.

If a molecule travels a distance of 1 cm, it sweeps out an imaginary volume of $\pi\sigma^2$ (1). With n molecules per unit volume the number of molecules struck per

Figure 3.3. Schematic of tunnel to estimate mean free path.

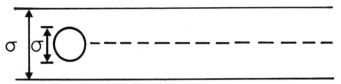

Table 3.1. Typical molecular diameters[a] [Bird et al., 1960]

Gas	σ,Å
H_2	2.9
N_2	3.6
O_2	3.4
Air (dry)	3.6

[a]R. B. Bird, et al., *Transport Phenomena,* Wiley, New York, 1960, p. 744.

centimeter is $\pi\sigma^2 n$, and the mean free path is then the reciprocal, $1/(n\,\pi\sigma^2)$. If it is assumed that the molecular velocities are distributed according to Maxwellian theory rather than having a single value, the mean free path equation is decreased by a factor of $\sqrt{2}$, or

$$\lambda = \frac{1}{\sqrt{2}\,n\pi\sigma^2} \qquad (3.12)$$

Typical molecular diameters for various gases are given in Table 3.1. The value given for air represents an average value considering the relative proportions of oxygen, nitrogen, and trace gases in standard dry air.

EXAMPLE 3.6

Find the mean free path of an "air" molecule at both 0°C and 20°C.

$$\lambda = \frac{1}{\sqrt{2} \times 2.69 \times 10^{19} \times \pi \times (3.6 \times 10^{-8})^2}$$

$$= 6.46 \times 10^{-6} \text{ cm.} = 0.065 \,\mu m$$

This is the mean distance between collisions at 0°C. At 25°C this distance would be 0.071 μm; at 20°C it would be 0.070 μm.

GAS VISCOSITY, HEAT CONDUCTIVITY, AND DIFFUSION

The three properties of viscosity, heat conductivity, and diffusion represent respectively the transfer of momentum, energy, and mass within a gas. The gas diffusion coefficient indicates the relative ability of one gas molecule to move with respect to its surroundings — the greater the value of the diffusion coefficient, the more rapid this movement. The diffusion coefficient, D, for a gas can be estimated from the expression

$$D = \frac{6\,\mu}{5\,\rho} \tag{3.13}$$

where μ is the gas viscosity and ρ the gas density (both in cgs units).

Viscosity can be estimated from the expression (Daniels and Alberty, 1961):

$$\mu = \frac{5\,[\pi m k T]^{0.5}}{16\,\pi\sigma^2} \tag{3.14}$$

The term m represents the mass of a single molecule. The equation does not include a pressure term or depend on molecular concentration. This is confirmed with real gases at moderate pressures and normal temperatures where the viscosity is essentially independent of pressure.

EXAMPLE 3.7

Determine the viscosity of helium gas at $20°C$.

Use 1.90 Å as the molecular diameter of helium.

Using Eq. 3.14,

$$\mu = \frac{5(\pi m k T)^{0.5}}{16\pi\sigma^2}$$

$$\mu = \frac{(5)[(3.14)(4/6.02 \times 10^{23})(1.38 \times 10^{-16})(273 + 20)]^{0.5}}{(16)(3.14)(1.90 \times 10^{-8})^2}$$

$$\mu = 113 \text{ micropoises}$$

Notice also that the viscosity varies as the square root of temperature. Thus, as the temperature goes up, the viscosity increases! This is just opposite to the behavior of typical liquids (for example, with motor oil the viscosity increases as the temperature decreases). For air at $20°C$ the viscosity, μ, is 1.83×10^{-4} poises.

For the aerosol scientist the main point to remember about the medium from a kinetic theory point of view is that mass, energy, and momentum can be transferred within the gas — mass by diffusion, energy by heat conduction, and momentum by viscosity.

Mean free path indicates the transfer of momentum, energy, or mass a distance of λ. In the steady state, the net transport equals zero. These forces, or transfer functions, always act to bring a system back to the steady state. This implies (1) diffusion from high concentration to low concentration; (2) heat conduction from hot to cold; and (3) momentum flow — mass motion energy to molecular motion (hence accompanied by a rise in temperature of the gas).

PROBLEMS

1. How many molecules of a gas are there per cm^3 at $20°C$; at $100°C$?

2. At $20°C$ the vapor pressure of water is 17.5 mm Hg. How many molecules of H_2O are there per cm^3 of air when the relative humidity is 50% and $T = 20°C$? How many ppm of water vapor?

3. Derive most probable gas molecule velocity.

4. Derive arithmetic mean gas molecule velocity.

5. Derive root-mean-square (rms) gas molecule velocity.

6. What is the magnitude of the rms velocity associated with a 0.1-μm diameter unit density spherical aerosol particle if it is in thermal equilibrium with its surroundings?

7. Compute the mean free path of an hydrogen molecule at $0°C$ using simple theory and then using a Maxwellian velocity distribution.

8. Using the equation given for viscosity, compute the viscosity of air at $20°C$.

FOUR

MACROSCOPIC FLUID PROPERTIES

REYNOLDS NUMBER

So far the properties of the medium have been discussed from a molecular point of view. Generally, however, the medium can be thought of as a continuum; that is, as a fluid where all molecules act in harmony with each other. This is the way one normally pictures a gas or liquid, and with this view of the medium the rules of aerodynamics can be applied.

Suppose it is desired to visualize the flow around a 1-μm sphere by studying the flow around a 1-cm sphere. One could ask, under what conditions is it reasonable to assume that a 1-μm diameter sphere moving in a continuous medium will behave in a manner similar to a 1-cm sphere moving in the same medium? Or more generally, under what conditions will geometrically similar flow occur around geometrically similar bodies? The answer, fundamental to fluid mechanics, is that in similar fields of flow, the forces acting on an element of either body must bear the same ratio to each other at any instant.

If the medium is considered incompressible and neglecting gravity, the main forces present are the inertial force due to the acceleration or deceleration of small fluid masses near the body and the viscous friction forces which arise due to the viscosity of the medium. For similarity these forces must be in the same ratio at any instant. Then

$$\frac{\text{Inertial force}}{\text{Viscous force}} = \frac{\rho_m \dfrac{v^2}{d}}{\mu \dfrac{v}{d^2}} = \frac{\rho_m v d}{\mu} = \text{Re} \qquad (4.1)$$

where v is the relative velocity between the fluid and the body, ρ_m is the density of the medium, μ is the medium viscosity and d is the body (or particle) diameter (Prandtl and Tietjens, 1934). The result is the *Reynold's number* (Re), a dimensionless number which describes the type of flow occurring around the body.

Kinematic viscosity, v, can be defined as

$$v = \frac{\mu}{\rho_m} \qquad (4.2)$$

then

$$\text{Re} = \frac{vd}{v} \qquad (4.3)$$

is a convenient form for computing Reynolds number in air at normal conditions. For air at normal pressure, $20°C$ temperature, the kinematic viscosity, v, is equal to 0.152 cm^2/sec.

EXAMPLE 4.1

A 1-in.-diameter sphere moves through air with a velocity of 10 in./min. Find its Reynolds number.

$$\text{Re} = \frac{vd}{v} = \frac{\dfrac{10 \times 2.54}{60} \times [1 \times 2.54]}{0.152} = 7.07$$

It is also possible to derive Reynolds number by dimensional analysis. This represents a more analytical, but less intuitive, approach to defining the condition of similar fluid flow and is essentially independent of particle shape. In this approach variables in the Navier-Stokes equation (relative particle-fluid velocity, a characteristic dimension of the particle, fluid density and fluid viscosity) are combined to yield a dimensionless expression. Thus

$$[v^\alpha \, d^\beta \, \rho_m^\gamma \, \mu^\delta] = F^0 \, L^0 \, T^0 = 1$$

Let $\alpha = 1$

$$\left(\frac{L}{T}\right)^1 (L)^\beta \left(\frac{FT^2}{L^4}\right)^\gamma \left(\frac{FT}{L^2}\right)^\delta = 1$$

Then

$$\gamma + \delta = 0$$

$$1 + \beta - 4\gamma - 2\delta = 0$$

$$2\gamma + \delta - 1 = 0$$

$$\beta = 1,$$

$$\gamma = 1,$$

$$\delta = -1$$

so

$$\mathrm{Re} = \frac{vd\rho_m}{\mu} \tag{4.1}$$

Table 4.1 gives typical values for viscosity, density, and kinematic viscosity for air at $0°C$ and $20°C$.

The Reynolds number is useful in describing the type of flow that is taking place. At high Reynolds numbers inertial forces will be much greater than viscous forces, while at low Reynolds numbers the opposite is true. Laminar or streamline flow is the result of the predominance of viscous forces. Thus at low Reynolds numbers the flow is laminar. Streamlines persist for great distances both upstream and downstream of the body and little mixing takes place. When inertial forces predominate, streamlines disappear and the flow is turbulent. With turbulent flow there is rapid and random mixing downstream of the body, and streamlines are relatively undisturbed in front of the body until they almost reach the body surface. In the range where Reynolds number increases from laminar flow to turbulence, the flow is said to be intermediate since at any time it can either be laminar or turbulent.

Laminar flow can also be known as Stokes flow or viscous flow while turbulent flow is sometimes termed potential flow.

Reynolds number can be applied to either a fluid flowing around a body or a fluid flowing inside a pipe. The transition from laminar to turbulent flow occurs at different Reynolds numbers for these two cases. Reynolds numbers at which different flow conditions prevail are tabulated in Table 4.2. Since v is the relative velocity between the medium and the body, Reynolds number is the same

Table 4.1.   Useful constants for air[a]

Property	$0°C$	$20°C$	Units
Viscosity	1.71×10^{-4}	1.83×10^{-4}	poises $=$ gm/(cm $\times$ sec)
Density	0.001293	0.001205	$=$ grams/cc
Kinematic viscosity	0.132	0.152	Stokes $=$ cm^2/sec

[a] P = 760 mm Hg

Table 4.2. Values of *Re* for various conditions of flow[a]

	A sphere of diameter d in a still fluid	Fluid flowing in a pipe of diameter d
Upper limit, laminar flow	1	2100
Intermediate region	1 – 1000	2100 – 4000
Turbulent flow	>1000	>4000

[a]It should be kept in mind that these values are approximate.

whether the body is moving through a stationary fluid or the fluid is flowing around a stationary body.

Schematic representations of these different flow conditions are illustrated in Fig. 4.1. As a gas enters a long pipe, turbulence will develop within the pipe if the Reynolds number exceeds the values given in Table 4.2.

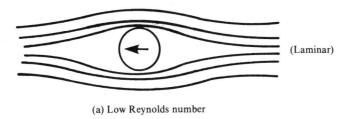

(Laminar)

(a) Low Reynolds number

(Intermediate)

(b) Intermediate Reynolds number

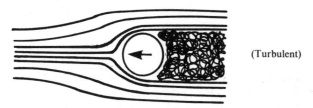

(Turbulent)

(c) High Reynolds number

Figure 4.1. Sketch of flow types.

Reynolds number is a fundamental parameter used to describe the fluid properties associated with an aerosol. Equations describing the resistance offered by a particle depend on whether the flow is laminar or turbulent, and the Reynolds number provides knowledge of the type of flow that is present.

EXAMPLE 4.2

An aerosol comprised of 1.0-μm-diameter spheres flows through a 16 in.-diameter duct with a velocity of 3500 fpm. Determine

    a. the Reynolds number of the air flowing in the duct
    b. the Reynolds number of the particles in the air.

    a. Reynolds number in duct $= \dfrac{(\text{duct diameter})\left(\substack{\text{relative velocity of air} \\ \text{in duct to duct}}\right)}{0.152}$

$$= \frac{(16 \times 2.54)\ 3500 \times \dfrac{30.5}{60}}{0.152}$$

$$= 4.76 \times 10^5$$

This is clearly turbulent flow.

    b. Reynolds number of particles $= \dfrac{(\text{particle diameter})\left(\substack{\text{relative velocity of} \\ \text{particles to air}}\right)}{0.152}$

$$= \frac{(1 \times 10^{-4})(0)}{0.152} = 0$$

Since the particles are moving at the same velocity as the air in the duct, their Reynolds number is zero.

DRAG

We can now consider the resistance offered by the medium to the motion of an aerosol particle. Some of the earliest interest in the motion of a body moving through a fluid arose from the desire to know where a cannon ball, once fired, would land.

This problem can be treated using the approach of Newton. Suppose the medium is composed of a large number of particles which have mass but no volume. These particles are everywhere at rest and are not connected. A body moving through this medium would experience impacts from the particles making up the medium and would inpart momentum to them. The mass of particles impacting per second on the body is $\rho_m A v$, where ρ_m is the density of the particles per unit volume (and thus also the density of the medium), A is the cross-sectional area of the body normal to the direction of motion, and v the body velocity. Each impacting particle is given some velocity v' on impact which is proportional to v. Thus the momentum "created" per second is $\rho_m A v v'$.

Since the time rate of change of momentum is a force, this is also equal to the resisting force of the medium to the motion of the particle, often called the *drag*, or

$$F_D = \rho_m A v v' = k \rho_m A v^2$$

where k is a constant. The momentum transferred to the medium actually depends on whether the impacts of the gas molecules on the body are elastic or inelastic. This is reflected in the value used for the constant k. Also, early estimates of k were incorrect because only the cross-sectional area of the body was considered, not the entire surface area. Depending on the type of flow, molecules can receive or impart momentum to the rear of the body, and the sides can have an influence so that the entire shape of the body is important, and not just its projected area.

There are three kinds of resistance which can be associated with the motion of a body as it passes through a medium. Deformation or viscosity drag represents the force necessary to deform the medium so that the body can pass through it. This deformation can occur at great distances both up and downstream of the body. A second source of drag is frictional resistance which occurs at the surface of the body. The third type of resistance, pressure drag, represents compression of the medium. These latter two types make up the so-called "skin friction" of the body. At small Reynolds numbers deformation drag predominates, and forces that act over the entire body surface must be taken into account. At large Reynolds numbers frictional resistance and pressure drag predominate. The drag in this case is primarily associated with the cross-sectional area normal to the fluid flow.

In cases involving high Reynolds numbers, Newton's approach (given above) agrees with experimental evidence, even though the underlying assumptions implying a constant value for k are wrong.

It is customary to write the drag equation as (Sutton, 1957)

$$F_D = (\text{some constant}) A \rho_m v^2$$

If a $[v^2/2]$ term (similar to the velocity head term in Bernoulli's equation) is used then

$$F_D = C_D A \rho_m (v^2/2)$$

where the constant C_D is now formally known as the *coefficient of drag*, or *drag coefficient*. For a sphere of diameter d,

$$A = (\pi/4) d^2$$

and then

$$F_D = [C_D \pi \rho_m d^2 v^2]/8 \qquad (4.4)$$

EXAMPLE 4.3

In turbulent flow the coefficient of drag is a constant with a value of about 0.4. What is the resisting force offered by air to a 6-in. cannon ball moving through the air with a velocity of 500 ft/sec.?

$$F_D = [C_D \pi \rho_m d^2 v^2]/8$$

$$= (0.4)(3.14)(0.0012)(6 \times 2.54)^2(500 \times 30.5)^2/8$$

$$F_D = 1.02 \times 10^7 \text{ dynes}$$

It was originally thought that for a given shape, body position, and relative velocity, C_D would be a constant. This is not the case, and it is not surprising in view of the many ways in which resistance to flow can arise, depending on Reynolds number. The coefficient of drag is a constant for a given shape and body position in those cases where the total drag is predominantly pressure drag (high Reynolds number). It is not a constant when deformation drag predominates (low Reynolds number).

Since we can have similarity of flow around similarly shaped bodies in those cases where the ratio of forces over the bodies' surfaces are the same, this is equivalent to saying that there will be similar resisting forces when the Reynolds numbers of the two bodies are the same. But then the drag coefficients for the two cases are also the same, i.e.,

$$C_D = f(\text{Re})$$

a statement which is true for each shape and body position. Fig. 4.2 shows the relationship of C_D versus Re for spheres. In some ranges of Reynolds numbers, C_D can be determined analytically. In others, it must be estimated empirically. For laminar flow (Re<1),

$$C_D = 24/\text{Re} \tag{4.5}$$

In the intermediate region (1<Re<1000) there ae many empirical formulas for C_D, e.g. (Crawford, 1976; Orr, 1966):

$$C_D = \frac{24}{\text{Re}} [1 + 0.15 \text{ Re}^{0.687}] \tag{4.6}$$

$$C_D = \frac{14}{\text{Re}^{0.5}} \quad \text{for } 2<\text{Re}<800 \tag{4.7}$$

$$C_D = \frac{24}{\text{Re}} + \frac{4}{\text{Re}^{0.33}} \tag{4.8}$$

$$C_D = \frac{18.5}{\text{Re}^{0.6}} \tag{4.9}$$

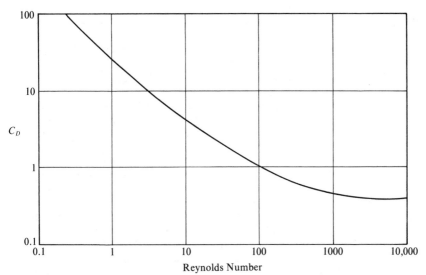

Figure 4.2. Coefficient of drag as a function of Reynolds number for spheres.

EXAMPLE 4.4

Compare values of C_D as computed from Eq. (4.6) through (4.9) for Re = 2. Which one is most nearly correct?

Estimated Values of C_D *from Equation*

Re		4.6	4.7	4.8	4.9
2	=	14.90	9.90	15.18	12.21

The reported measured value for C_D for spheres with Re = 2 is 14.6. Equations (4.7) and (4.9) are not very accurate, but they can be useful because of their simplicity. Where greater accuracy is needed, Eq. 4.6 or 4.8 should be used.

In the lower turbulence region ($1000 < $ Re $ < 2 \times 10^5$)

$$C_D = 0.44 \qquad (4.10)$$

and in the upper turbulence region (Re $> 2 \times 10^5$)

$$C_D = 0.10 \qquad (4.11)$$

Table 4.3 gives computed values for C_D for various values of Re based on these different equations, compared to actual measurements of drag coefficients for spheres.

Table 4.3. Experimental and computed values of C_D as a function of Re[a]

Re	Experimental	(4-6)	(4-7)	(4-8)	(4-9)
			Approximations		
0.1	240	247.4	44.3	248.6	73.6
0.2	120	126.0	31.3	126.8	48.6
0.3	80	85.3	25.6	86.0	38.1
0.5	49.5	52.5	19.8	53.0	28.0
0.7	36.5	38.3	16.7	38.8	22.9
1.0	26.5	27.6	14.0	28.0	18.5
2	14.6	14.9	9.9	15.2	12.2
3	10.4	10.6	8.1	10.8	9.57
5	6.9	7.0	6.3	7.14	7.04
7	5.3	5.4	5.3	5.52	5.76
10	4.1	4.2	4.43	4.26	4.7
20	2.55	2.61	3.13	2.67	3.07
30	2.00	2.04	2.56	2.09	2.40
50	1.50	1.54	1.98	1.57	1.77
70	1.27	1.30	1.69	1.31	1.45
100	1.07	1.09	1.40	1.10	1.17
200	0.77	0.81	0.99	0.80	0.77
300	0.65	0.68	0.81	0.68	0.60
500	0.55	0.56	0.63	0.55	0.44
700	0.50	0.50	0.53	0.48	0.36
1000	0.46	0.44	0.44	0.42	0.29
2000	0.42	0.35	0.31	0.33	0.19
3000	0.40	0.30	0.26	0.29	0.15
5000	0.385	0.26	0.20	0.24	0.11
7000	0.390	0.23	0.17	0.21	0.09
10,000	0.405	0.20	0.14	0.19	0.07
20,000	0.45	0.16	0.10	0.15	0.05
30,000	0.47	0.14	0.08	0.13	0.04
50,000	0.49	0.12	0.06	0.11	0.03
70,000	0.50	0.11	0.05	0.10	0.02

[a]R. H. Perry and C. H. Chilton, *Chemical Engineers Handbook,* 5th Ed., McGraw-Hill, New York, 1973, pp. 5-64.

EXAMPLE 4.5

A particle of diameter d and density ρ settles under the influence of gravity.

What is its terminal settling velocity?

For terminal settling the drag force equals the force due to gravity. Hence

$$F_D = F_G$$

$$mg = C_D A \rho_m \frac{v^2}{2}$$

for spheres

$$\frac{\pi}{6} d^3 (\rho_p - \rho_m) g = C_D \frac{\pi}{4} \rho_m \frac{v^2}{2}$$

$$v^2 = \frac{4d(\rho_p - \rho_m)g}{3 C_D \rho_m}$$

Unfortunately, C_D depends on Re which depends on v. However,

$$C_D Re^2 = C_D \frac{v^2 d^2 \rho_m^2}{\mu^2}$$

so substituting for v^2 gives

$$C_D Re^2 = C_D \frac{d^2 \rho_m^2}{\mu^2} \left[\frac{4}{3} \frac{d^3 \rho_m (\rho_p - \rho_m)g}{3 C_D \rho_m} \right]$$

$$C_D Re^2 = \frac{4}{3} \frac{d^3 \rho_m (\rho_p - \rho_m)g}{\mu^2} \tag{4.12}$$

$C_D Re^2$ can be computed from Eq. (4.12) since the v term has been eliminated and from a plot of $C_D Re^2$ versus Re (Fig. 4.3) a value of Re can be found which yields v. This method, although crude, is valid for determining settling velocities for any size particle in either a gas or a liquid. For most aerosol particles a simpler method is available, as will be discussed in the next chapter.

EXAMPLE 4.6

Determine the settling velocity of a 100-μm-diameter gold sphere (ρ = 19.3 g/cm^3)

when it settles in

   a. air
   b. water.

   a. The density of air is 0.0012 g/cm^3, so using Eq. 4.12 gives

$$C_D Re^2 = \frac{4}{3} \frac{d^3 \rho_m (\rho_p - \rho_m)g}{\mu^2}$$

$$C_D Re^2 = \frac{4}{3} \frac{(100 \times 10^{-4})^3 (0.0012)(19.3 - 0.0012)(980)}{(1.83 \times 10^{-4})^2}$$

$$C_D Re^2 = 904$$

From Figure 4.3, $C_D\text{Re}^2 = 904$ gives a value of Re = 17.8. Hence

$$v = \frac{\text{Re}v}{d} = (17.8)(0.152)/(100 \times 10^{-4}) = 271 \text{ cm/sec}$$

b. For water, $\rho_m = 1$ and $\mu = 0.01$ so that

$$C_D\text{Re}^2 = \frac{4}{3}\frac{(100 \times 10^{-4})^3(1)(19.3 - 1)(980)}{(0.01)^2}$$

$$C_D\text{Re}^2 = 239$$

Again from Figure 4.3, $C_D\text{Re}^2 = 239$ gives a value of Re = 6.31. Hence

$$v = \frac{\text{Re}v}{d} = (6.31)(0.01)/(1)(100 \times 10^{-4}) = 6.31 \text{ cm/sec}$$

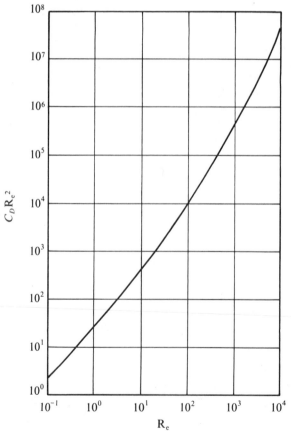

Figure 4.3. Plot of $C_D\text{Re}^2$ versus Re.

EXAMPLE 4.7

Given particle density and settling velocity, how can particle diameter be determined for a settling particle?

Again equating forces,

$$F_G = F_D$$

$$mg = C_D A \rho_m \frac{v^2}{2}$$

so that for spheres

$$d = \frac{3v^2 C_D \rho_m}{4\rho_p g}$$

The term C_D can be eliminated by taking the ratio

$$\frac{C_D}{Re} = \frac{C_D \mu}{dv\rho_m} = \frac{C_D \mu 4 \rho_p g}{3v^3 C_D \rho_m^2}$$

$$\frac{C_D}{Re} = \frac{4\mu\rho_p g}{3v^3 \rho_m^2} \tag{4.13}$$

Here C_D/Re can be computed from Eq. (4.13). Using Fig. 4.4 (a plot of C_D/Re vs. Re) the Re can be found, which then yields d. For small particles settling in the Stokes region, this involved process is not necessary.

PROBLEMS

1. Calculate the density of CO_2 at $20°C$.

2. Calculate the Reynolds number of a 1-μm spherical sand particle moving in air at a velocity of 10 c/sec (assume NTP).

3. Calculate the Reynolds number of a 10-in. ball moving in air at a velocity of 10 cm/sec. Is the flow around the ball laminar or turbulent?

4. Calculate the Reynolds number for air flowing through a 10-in. diameter pipe at a velocity of 10 cm/sec. Is the flow through the duct laminar or turbulent?

5. A 10-μm-diameter particle settles in air with a velocity of 0.30 cm/sec. If the settling of this particle is to be modeled by a one inch diameter steel ball moving in glycerol (viscosity = 1756 centipoises), what should be the ball's velocity in the glycerol? (Density of glycerol = 1.26 g/cm³.)

6. Air flows through a 4-in.-diameter duct at a rate of 100 cfm. Determine whether this flow is laminar or turbulent within the duct.

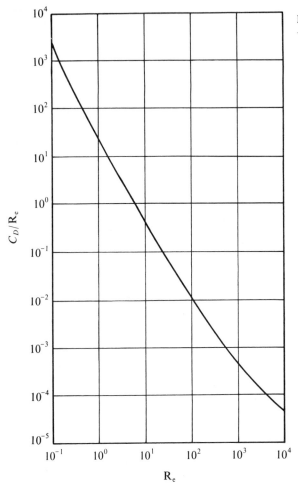

Figure 4.4. Plot of $C_D/$Re versus Re.

7. A 10-μm-diameter particle falls in still air with a velocity of 0.30 cm/sec. If the drag coefficient is given by 24/Re, what is the force developed by the falling particle.

8. Using Eq. (4.8), calculate C_D, C_DRe^2 and $C_D/$Re for the following values of Re:
   a. 0.8
   b. 80
   c. 8000

9. Determine, using a plot of $C_D/$Re and C_DRe^2 vs. Re:
   a. the settling velocity of a 200-μm sand sphere ($\rho = 2.65$ g/cm^3)
   b. the size of a water droplet that settles at a velocity of 10 cm/sec.

VISCOUS MOTION, STOKES' LAW

INTRODUCTION

For the case of low Reynolds number flow (viscous flow), it is possible to develop an expression for the force resisting the motion of a sphere moving through a fluid based purely on mathematical reasoning. This problem was originally solved by G. G. Stokes and the expression for force since has become known as "Stokes' law". For those interested in the mathematics of this problem, Stokes' derivation is given in Appendix B. Although one doesn't have to understand the derivation to use Stokes' law correctly, it helps to be aware of the assumptions that were made in order to understand how deviations from these assumptions can affect results of calculations made using Stokes' law.

For Stokes' solution, it was necessary to assume a continuous, incompressible, viscous, and infinite medium with rigid particles and spherical particle shapes. With these assumptions, Stokes found that the resisting force exerted by air on a moving particle, equivalent to the force exerted by moving air on a stationary particle, is:

$$F = 3\pi\mu vd \qquad (5.1)$$

where F is the force on the particle, in dynes, μ is the viscosity of the medium in poises, v is the relative velocity between the air and the particle in centimeters per second and d is the diameter of the sphere in centimeters.

51

The best proof of the validity of Stokes law (although indirect) was the Millikan Oil Drop experiment. Stokes law has been shown to give a reasonable approximation of the resisting force on spheres in many other situations, and the only stipulation is that the assumptions listed above are not violated.

EXAMPLE 5.1

A 1-μm unit density sphere moves through air with a velocity of 100 cm/sec. Compute the magnitude of the resisting force offered by the air. Assume $T = 20°$ C, 760 mm atmospheric pressure.

$$F = 3 \pi \mu v d$$

$$F = [3] [3.14] [1.83 \times 10^{-4}] [100] [1 \times 10^{-4}]$$

$$F = [gm/(cm\text{-}sec)] [cm/sec] [cm] = [gm\text{-}cm/sec]$$

$$F = 1.72 \times 10^{-5} \text{ dynes}$$

Stokes' law becomes incorrect when assumptions used to derive it cannot be met. It is possible in some cases to develop correction factors broadening the conditions under which Stokes' law is applicable. On the other hand, the assumptions may be so broad for the types of problems which are of interest that corrections are not necessary or are impossible to make. In any case, it is useful to examine each assumption in detail to determine when it may or may not be valid.

CONTINUOUS MEDIUM

When the diameter of a particle is very small, approaching the mean free path of the molecules in the medium, Cunningham (1910) and also Millikan (1910) showed that because the medium is no longer a "perfect" continuum the resisting force offered to the particle should be smaller than predicted by Stokes' law. The difference in the dependence of resistance on particle diameter corresponds to the conditions prevailing at the two extreme particle ranges. For large particles the primary source of resistance is the viscosity of the medium, whereas with small particles or with a highly rarified medium viscosity is no longer important and the predominant resisting mechanism is due to the inertia of the gas molecules which the particle encounters. As particle size decreases to near molecular size, the resisting force offered by the medium becomes a function of the cross-sectional area of the particle, consistent with Newton's model for drag (Millikan, 1923).

To correct for this effect a factor, commonly known as the *Cunningham correction factor, slip,* or *Millikan resistance factor,* and denoted C_c, must be introduced into the Stokes equation, yielding

$$F = [3\pi\mu v d]/C_c,$$

where   (5.2)

$$C_c = 1 + \frac{2\lambda}{d}\left[A + Q \exp\left(-\frac{bd}{2\lambda}\right)\right]$$ (5.3)

In this equation $A = 1.257$, $Q = .400$, $b = 1.10$, and λ is the mean free path of the gas molecules.

The Cunningham correction factor, C_c, is always equal to or greater than one.

When $d > 2\lambda$, C_c can be approximated by the expression

$$C_c = \left[1 + \frac{2\lambda}{d}(1.257)\right].$$ (5.4)

When $d < 2\lambda$, then C_c becomes approximately equal to

$$C_c \cong \left[1 + \frac{2\lambda}{d}(1.657)\right].$$ (5.5)

The Cunningham correction factor is an important correction to Stokes' law and should always be used when particles are less than 1 μm in diameter.

EXAMPLE 5.1

Compute the Cunningham correction factor for a silica dust particle ($\rho = 2.65$ g/cm^3) having a diameter of 0.5 μm. Assume a spherical shape, and 20°C.

From Eq. (5.3)

$$C_c = \left(2 + \frac{2\lambda}{d}\left[A + Q \exp\left(-\frac{bd}{2\lambda}\right)\right]\right)$$

The gas mean free path (from Chapter 3) is 0.070 μm, so

$$C_c = \left(1 + \frac{(2)(0.070)}{(0.5)}\left[1.257 + 0.4 \exp\left(-\frac{(1.10)(0.5)}{2(0.070)}\right)\right]\right)$$

$$C_c = 1.35$$

Table 5.1 gives values of C_c for various particle diameters.

As mentioned earlier, the slip, or Cunningham correction factor, represents the mechanism for transition from the continuum to the molecular case. For large values of d, F is proportional to d whereas for small values of d, F is proportional to the square of d. Figure 5.1 shows a plot of C_c as a function of particle diameter for air at normal conditions of temperature and pressure.

Table 5.1. Computed values of C_c for various diameters

Diameter, μm	C_c
10.0 μ	1.018
1.0 μ	1.176
0.1 μ	3.015
0.01 μ	23.775
0.001 μ	232.54

INCOMPRESSIBLE MEDIUM

Air is compressible, but compression is not important for motion in the Stokes' region. This assumption can be considered to be always valid.

VISCOUS MEDIUM

In the derivation of Stokes' law, the assumption of a perfectly viscous medium means that no inertial forces are considered. This was done to linearize the Navier-Stokes' equation. If these inertial effects are included using a first-order approximation, it is possible to extend the applicability of Stokes' law up to a

Figure 5.1. Plot of C_c as a function of d, normal temperature and pressure.

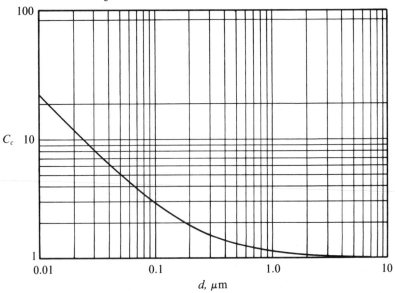

Reynolds number of about 5. Then the resisting force can be expressed as

$$F = 3\pi\mu vd \ (1 + [3/16] \ \text{Re}) \tag{5.6}$$

Above a Reynolds number of about 5, Stokes' law, even with this correction, is no longer applicable.

EXAMPLE 5.2

A 100-μm unit density sphere moves through air with a velocity of 30 cm/sec. Compute the resisting force offered by the air, in dynes. Assume normal temperature and pressure

$$\text{Re} = \frac{dV}{v} = \frac{(100 \times 10^{-4})(30)}{0.152} = 1.97$$

$$F = 3\pi\mu vd \ \left(1 + \frac{3}{16} \ \text{Re}\right)$$

$$= 3(3.14)(1.83 \times 10^{-4})(30)(100 \times 10^{-4}) \left[1 + \frac{3}{16}(1.97)\right]$$

$$= 7.09 \times 10^{-4} \ \text{dynes}$$

INFINITE MEDIUM

In viscous flow perturbations caused by a particle extend large distances into the medium. The presence of other particles moving nearby will have the effect of reducing the resistance of the medium to that particle by setting the medium near the particle in motion. Hence an ensemble of particles will settle faster than they would as isolated entities, and when two equal sized particles fall along the same axis, the upper of the two will fall faster that the lower, so that they will eventually collide. If the particles are of different diameters, the aerodynamic interaction between the two particles will result in an increase in settling velocity for both particles. When the leading particle is smaller than the trailing particle, its increase in velocity will be greater than the increase for the trailing one. Particle induced interactions are usually neglected in making estimates of settling rates since with the exception of the most extreme cases particle-particle spacing is relatively large.

EXAMPLE 5.3

Typical concentrations for condensation nuclei are 30,000 to 50,000 nuclei/cm^3. If each nuclei is 0.01 μm in diameter, and particles are present in a concentration of 40,000/cm^3 estimate average particle spacing, in particle diameters.

$$\text{cm}^3/\text{particle} = \frac{1}{40,000}$$

$$\text{cm}/\text{particle} = \sqrt[3]{\frac{1}{40,000}}$$

$$= 0.029$$

$$= \frac{0.029}{0.01 \times 10^{-4}} = 29,240 \text{ particle diameters}$$

This spacing is sufficiently large so that particle-particle interactions can be neglected.

When particles move parallel to a flat surface resistance is increased due to the drag induced by the surface. This increase is so small, and extends such a small distance into the medium (several particle diameters at most), that the effect can be neglected without significant error.

For aerosols in a confined space other interaction effects are possible. For example, a cloud of sedimenting particles could completely fill a finite volume. Then the downward motion of each single particle creates a downward flow field that tends to pull along neighboring particles. But in a confined space this downward flow is balanced by an upward air flow that tends to lift the entire cloud. The net result is that the downward velocity of the cloud in a finite container will be less than that of a similar cloud in an infinite medium. Figure 5.2 illustrates the two cases of confined and unconfined aerosol sedimentation. For a complete discussion of non-infinite medium effects, see Happel and Brenner (1965).

RIGID PARTICLES

Although rigid particles are assumed, often Stokes' law is applied to nonrigid or liquid droplets. In the case where the drops are large, they are deformed by the motion of the air and will no longer be spherical. Since they tend to flatten out they offer more resistance to falling, and have lower terminal velocities than spherical particles. This effect is not important for freely falling particles having diameters less than a few hundred micrometers. Table 5.2 gives terminal settling velocity data for raindrops of various diameters. Above about 6 mm in diameter the drops fracture and break up while falling.

More importantly, nonrigid particles can undergo internal circulation as they move through a medium. This circulation reduces the friction at the drop surface

Figure 5.2. Sketch of generalized flow patterns for cases of unconfined and confined settling.

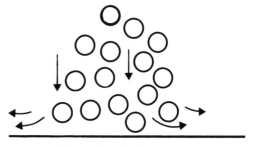

Unconfined Sedimentation

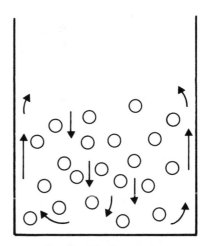

Confined Sedimentation

so that the resistance offered by the medium to the motion of the drop is reduced. The resisting force then becomes

$$F_c = 3\pi\mu_m vd \left[\frac{1 + \dfrac{2\mu_m}{3\mu_p}}{1 + \dfrac{\mu_m}{\mu_p}} \right]$$

(5.7)

where μ_p is the viscosity of the liquid making up the drop and μ_m is the viscosity of the medium. For water droplets in air the correction factor is for all practical purposes equal to one, since the viscosity of water is so much greater than that of air. In general, for liquids in air this effect can be neglected.

Table 5.2 Terminal velocities of water droplets in still air (NTP)[a]

Drop Diameter (μm)	V_T (cm/sec)
100	25.6
120	34.5
160	52.5
200	71
400	160
600	246
800	325
1000	403
2000	649
3000	806
4000	883
5000	909
5400	914
5800	917

[a]B. J. Mason, *The Physics of Clouds,* 2nd Ed., Clarendon Press, Oxford, 1971, p. 594.

EXAMPLE 5.4

Compare the resisting force of air on a 1-μm water droplet falling freely if the liquid nature of the droplet is considered.

Resistance allowing the liquid nature of droplet

$$F_C = 3\pi\mu_m vd \left[\frac{1 + (2\mu_m/3\mu_p)}{1 + (\mu_m/\mu_p)} \right]$$

Resistance neglecting liquid nature of droplet

$$F = 3\pi\mu_m vd$$

Taking the ratio gives

$$\frac{F_C}{F} = \frac{1 + (2\mu_m/3\mu_p)}{1 + (\mu_m/\mu_p)}$$

If $\mu_m = 1.83 \times 10^{-4}$ poise and $\mu_p = 1$ poise

$$\frac{F_C}{F} = \frac{1 + (2 \times 1.83 \times 10^{-4}/3)}{1 + (1.83 \times 10^{-4})/1)}$$

$$\frac{F_C}{F} = \frac{1.000122}{1.000183} = 0.999939$$

It is clear that this effect can be neglected.

If the viscosity of the medium greatly exceeds that of the droplet, the correction factor tends to a limiting value of two-thirds. In this case the resisting force becomes

$$F = 2\pi\mu_m vd \tag{5.8}$$

which is the resisting force a liquid offers to a bubble rising through it.

EXAMPLE 5.5

How fast will a 0.1-mm bubble rise in a glass of beer?

$$F_R = F_G = 2\pi\mu_m vd = mg$$

$$v = \frac{\dfrac{\pi}{6}d^3(\rho_p - \rho_m)g}{2\pi\mu_m d} = \frac{1}{12}\frac{d^2(\rho_p - \rho_m)g}{\mu_m} = \frac{10^{-4}(-1)(980)}{12(1)}$$

$$v = 0.817 \text{ cm/sec}$$

Check:

$$Re = \frac{dv\rho}{\mu} = \frac{(0.01)(0.817)(1)}{0.01} = 0.817$$

This indicates laminar flow, so assumptions are O.K.

When the viscosities of the medium and the particle are the same, then the correction factor has a value of $(5/6)$, and the Stokes' resistance is

$$F = (5/2)\pi\mu_m vd \tag{5.9}$$

equivalent to the case of a cloud of particles being considered as a single particle having the same viscosity as the air. In this case d is the diameter of the entire cloud.

EXAMPLE 5.6

Wind blowing on an aerosol can either move it as a cloud or blow through it and dissipate it. Find the concentration of an aerosol made up of 5-μm water droplets which will be just dissipated by the wind.

Force on an individual particle:

$$F_S = 3\pi\mu_m vd$$

Force on an ensemble of particles (assuming spherical cloud of 10 meters diameter, D):

$$F_E = \frac{5}{2}\pi\mu_m vD$$

Since resisting forces tend to a minimum value, if the sum of the forces acting on all the particles is greater than the single force acting on the ensemble of particles, the particles will remain as an aerosol cloud. Otherwise, the cloud will dissipate.

If c is the aerosol concentration (p/cm^3), then the point where the forces are just equal is

$$\frac{5}{2}\pi\mu_m vD = 3\pi\mu_m dv[-c(\pi/6)D^3]$$

solving for c gives

$$c = \frac{5}{\pi dD^2}$$

with $d = 5$ μm, and $D = 10$ m,

$$c = \frac{5}{\pi \times 5 \times 10^{-4} \times 10^6} = \frac{1}{\pi} = 3.18 \times 10^{-3} \; p/cm^3$$

Concentrations greater than this value will result in the cloud remaining intact. This indicates that it is quite difficult to dissipate a cloud without some external aid other than mere blowing.

SPHERICAL PARTICLE SHAPE

A final assumption made in the derivation of Stokes' law was that the particles of interest were spheres. In many cases this is not true. Particles may have irregular shapes, depending on how they were formed and the amount of agglomeration which may have taken place. Liquid aerosols are always spherical in shape, so that for liquid aerosols the assumption of sphericity holds. For isometric particles this assumption can also be used with little error. For long chains of particles or flocculated particles, large deviations from Stokes' law are possible.

To use Stokes' law with chains or fibers, a correction factor, κ, known as the dynamic shape factor, is defined such that

$$F = 3\pi\mu vd_e\kappa \tag{5.10}$$

The term d_e is the diameter of a sphere having the same volume as the chain or fiber, that is

$$\text{Volume (chain or fiber)} = \frac{\pi}{6} d_e{}^3 \qquad (5.11)$$

For a cluster of n spheres of diameter d, $d_e = \sqrt[3]{n}\, d$. When the aggregate particle size is small the Cunningham correction factor should be considered.

Quite good estimates of the numerical value of the term κ have been made experimentally (Stöber and Flachsbart, 1969), and Table 5.3 shows some of these data. For tightly packed clusters, the maximum value for κ is about 1.25.

Table 5.3. Values of κ for different chain configurations[a]

n	Configuration	κ
2	oo	1.12
3	ooo	1.27
3	o o / o	1.16
4	oooo	1.32
4	oo with o above and o below	1.25
5	ooooo	1.45
6	oooooo	1.57
4	oo / oo	1.17
7	ooooooo	1.67
5	ooo with o, o	1.30
6	oooo with o, o	1.43
8	oooooooo	1.73
8	oooooo / o / o	1.64
5	o o / o / o o	1.19
8	oooooo with o, o	1.56
6	oo / oo / oo	1.17

[a]Adapted from Stöber and Flachsbart, *EST, 3,* 1280 (1969).

EXAMPLE 5.7

Determine the aerodynamic diameter of a particle made up of four spheres of 10-μm diameter (unit density) and formed into a tight cluster.

By equating forces, $F_R = F_G$,

$$v_T = \frac{mg}{3\pi\mu d_e\kappa} = \frac{(4)\left(\frac{\pi}{6}d^3\right)\rho_p g}{3\pi\mu(\sqrt[3]{4}d)\kappa}$$

Aerodynamic diameter, d_A, can be defined as

$$d_A{}^2 = \frac{18 v_T \mu}{g}$$

Hence

$$d_A{}^2 = \frac{18(4)\left(\frac{\pi}{6}d^3\right)\rho_p g\mu}{3\pi\mu(\sqrt[3]{4}d)\kappa g} = \frac{4d^2\rho_p}{\sqrt[3]{4}\kappa}$$

$$d_A = \left[\frac{\rho_p}{\kappa}\right]^{1/2}\sqrt[3]{4}d$$

For unit density spheres, $\rho = 1$, so that

$$d_A = \frac{\sqrt[3]{4}d}{\sqrt{\kappa}} = \frac{\sqrt[3]{4}}{\sqrt{1.17}}d = 1.468\,d = 14.68\ \mu m$$

With fibers, measurements are usually in terms of fiber length, L, and diameter d. Writing aerodynamic diameter as

$$d_A{}^2 = \frac{18\,v_T\mu}{g} \tag{5.12}$$

and replacing for v_T an expression derived from equating gravitational and resisting forces,

$$v_T = \frac{(\pi/4)d^2 L\rho_p g}{3\pi\mu\left[\frac{3}{2}d^2 L\right]^{1/3}\kappa} \tag{5.13}$$

yields

$$d_A = \left(\frac{3}{2}\right)^{1/3}\left(\frac{\rho_p}{\kappa}\right)^{1/2}\left(\frac{L}{d}\right)^{1/3}d \tag{5.14}$$

For chainlike aggregates a good approximation of κ can be made by assuming that (Stöber, 1972)

$$\kappa \times \frac{\pi d_e^2}{A} = \text{constant}$$

where A is the surface area of the particle. This implies that κ is proportional to $(L/d)^{1/3}$. Thus

$$d_A = \phi \rho^{1/2} \left(\frac{L}{d}\right)^{1/6} d \qquad (5.15)$$

where ϕ is a constant having a value for chainlike aggregates approximately equal to 1.08 $(\text{cm}^3/\text{gm})^{1/2}$. For asbestos fibers, assuming a density of 2.65 gm/cm^3, ϕ is approximately 1.35 $(\text{cm}^3/\text{gm})^{1/2}$ (Spurny et al., 1978).

EXAMPLE 5.8

Estimate the aerodynamic diameter of an asbestos fiber having a length of 15 μm, a diameter of 0.4 μm and a density of 2.65 g/cm^3.

From Eq. (5.15),

$$d_A = \phi \rho^{1/2} \left(\frac{L}{d}\right)^{1/6} d = 1.35 \sqrt{2.65} \left(\frac{15}{0.4}\right)^{1/6} (0.4)$$

$$d_A = 1.61 \ \mu\text{m}$$

Eq. (5.15) implies that the aerodynamic diameter of a rod or fiber will be influenced very little by its length, being much more dependent on its cross-sectional diameter. Hence fibers of different lengths but similar cross-sections will have similar aerodynamic properties, despite possibly large differences in mass.

PROBLEMS

1. Compare the force resisting the movement of a 100-μm-diameter sphere as it moves through air at a velocity of 1 cm/sec to a sphere moving in water at the same velocity.

2. Compute the force on a 10-μm unit density sphere as it settles at a velocity of 0.3 cm/sec.

3. At a Re of 4, the measured value of C_D for a sphere is 8.472. Determine the error in using Stokes' law with and without the appropriate correction factor.

4. Determine an expression for the force resisting the movement of an air bubble in water (assume Stokes' law holds). Find the value of C_D for such a system when Re = 1.

5. Compute the value of C_c for a 0.5-μm-diameter sphere:
   a. in air at 20°C, 760 mm pressure
   b. in air at 0°C, 0.25 atm pressure

6. Show, using Stokes' law and the slip correction factor, that for very small particles the resisting force is proportional to d^2.

7. Given a particle made up of a 2-sphere cluster, each sphere having a density of 2 gm/cm^3 and a diameter of 1 μm, find the aerodynamic and Stokes diameter of the cluster.

8. What length 1-μm diameter fiber will have the same aerodynamic diameter as a 10-μm unit density sphere? Assume the fiber density is 2.65 g/cm^3.

PARTICLE KINETICS
Settling, Acceleration, Deceleration

Kinetics is the study of changes in particle motion due to various forces acting on the particle. Particle motion can be rectilinear, that is, along a straight line, with the particle perhaps accelerating or decelerating as it moves, or the motion can be curvilinear, caused by forces acting to make a particle change its direction of motion. Rectilinear motion is covered in this chapter, curvilinear motion in the next.

When a force is applied to a particle at rest it begins to accelerate. If the particle is in air, the force of the air resisting the motion is zero when the particle is at rest but increases as the motion of the particle increases. As discussed in Chapter 5, at low Reynolds numbers this resisting force is given by Stokes' law. If the accelerating force is constant, eventually a point will be reached where the resisting force and accelerating forces are equal and the particle will then move at a constant velocity. Often it is important to know how long it will take before a particle starting from rest will reach this constant velocity, or if it has some initial velocity, how long it will take to reach its final velocity. This is important in determining the velocity necessary for capture of particles by a ventilation system, and is of interest in determining how quickly particles attain a constant, or terminal settling velocity after they are dropped.

EXAMPLE 6.1

Carbon particles in the exhaust of a diesel truck travelling at 55 mph are discharged into the atmosphere. Assuming that as these particles leave the exhaust they have the same velocity as the truck, how far will they travel in air before their motion is essentially stopped?

For all practical purposes, they lose their initial velocity immediately on discharge. This will become apparent later in the chapter.

EQUATION OF MOTION OF AN AEROSOL PARTICLE

By determining the path of a single particle when it is acted upon by a variety of forces it is possible to predict particle position and behavior. This can be done by solving force balance equations which then give acceleration, velocity, and position of the particle.

The net difference of the forces acting on the particle is equal to the rate of change of particle momentum. Thus

$$m \frac{d\vec{v}}{dt} = \vec{F}_1 + \vec{F}_2 + \vec{F}_3 + \vec{F}_4 + \dots \qquad (6.1)$$

where m is the mass of the particle. The forces $\vec{F}_1$, $\vec{F}_2$, etc. may include those which are generally functions of time and the position of the particle, such as electrical or magnetic forces, or they can be forces which are constant, such as gravity. These forces are generally balanced against the drag force, which depends on the properties of the medium, the field of flow, particle shape, and the instantaneous particle velocity. For many aerosol problems, this force is taken to be equal to the Stokes resistance, $3\pi\mu v d$, with the appropriate corrections, and it always acts in a direction opposite to the instantaneous particle motion.

If all forces are balanced, that is, $m\, dv/dt = 0$, the particle is not accelerating and moves with a uniform velocity if it moves at all. When the Stokes resistance is equal to zero, the particle velocity with respect to the airstream is zero.

Equation (6.1) represents a system of three second-order ordinary differential equations for the coordinates x, y, and z (or for some curvilinear coordinates q_1, q_2, q_3) expressed as functions of the time, t. Solution of these equations defines a trajectory of the particle for certain initial conditions of position and velocity. We shall examine several examples.

EXAMPLE 6.2

Write a force balance equation for a particle which is acted upon by gravity in an electric field.

Since the direction of motion of the particle is unspecified, our equation must be flexible in terms of direction. This is done by writing the equation in vector notation (the arrow over the variable indicates that the variable is a vector, it has both magnitude and direction).

Electrical force is given by

$$\vec{F}_E = q\vec{E}$$

where q is the charge on the particle and $\vec{E}$ is the field strength (a vector quantity).

The gravitational force is given by

$$\vec{F}_G = mg\vec{G}$$

Here m is the mass of the particle, g the acceleration due to gravity (both non-vector quantities), and $\vec{G}$ is a unit vector which establishes the direction in which gravity acts.

Then (since velocity is also a vector quantity)

$$m\frac{d\vec{v}_1}{dt} = mg\vec{G} + q\vec{E} + 3\pi\mu d\vec{v}_2$$

Notice that each set of terms in the equation contains one vector quantity (i.e., each term specifies a direction as well as a magnitude). Also notice that $\vec{v}_1$ represents the absolute particle velocity whereas $\vec{v}_2$ is the particle velocity relative to the medium velocity. Thus if $\vec{u}$ is the medium velocity, $\vec{v}_2 = \vec{u} - \vec{v}_1$.

PARTICLE MOTION IN THE ABSENCE OF EXTERNAL FORCES EXCEPT GRAVITY

Consider the case of a spherical aerosol particle moving in a homogeneous air-stream with no forces acting on the particle except gravity. For simplicity the motion will be assumed to occur only in the Stokes region (in most cases this assumption is valid). Then (similar to Ex. 6.2),

$$m\frac{d\vec{v}}{dt} = 3\pi\mu d(\vec{u} - \vec{v}) + mg\vec{G} \qquad (6.2)$$

where m is the particle mass, $\vec{v}$ the velocity of the center of gravity of the aerosol particle, $\vec{u}$ the velocity of the airstream near the particle, and $\vec{G}$ is the unit vector of the force of gravity. Dividing by $3\pi\mu d$ and rearranging terms gives

$$\tau\frac{d\vec{v}}{dt} + \vec{v} = \vec{u} + \tau g\vec{G} \qquad (6.3)$$

where

$$\tau = \frac{m}{3\pi\mu d} = \frac{m}{(F_R/v)} \tag{6.4}$$

The factor τ is an extremely important parameter in aerosol studies, as will be shown later. Properties of the particle (diameter and density) and of the medium (viscosity and density) are incorporated in this parameter, which is in units of seconds. It represents a relaxation time for the aerosol particle.

For spherical particles of mass m, $m = \frac{\pi}{6}d^3(\rho_p - \rho_m)$, τ becomes

$$\tau = \frac{1}{18}\frac{d^2(\rho_p - \rho_m)}{\mu}$$

Since for air $\rho_p \gg \rho_m$, τ is usually written as

$$\tau = \frac{1}{18}\frac{d^2}{\mu}\rho_p \tag{6.5}$$

The terms $\vec{u}$ and $\tau g\vec{G}$ represent two constant vectors which can be added together to form a single constant vector, $\vec{u}_0$. This addition is shown schematically in Fig. 6.1.

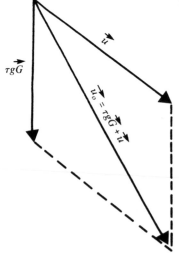

Figure 6.1.   Vector diagram showing definition of u_0.

EXAMPLE 6.3

Air flows in a horizontal duct with a velocity of 4 cm/sec. If the acceleration due to gravity is 980 cm/sec^2, determine the numerical value of the constant vector $\vec{u}_0$ (for a 30-μm diameter particle $\tau = 2.73 \times 10^{-3}$ sec).

$$(u_o)^2 = (u)^2 + (\tau g)^2 + 2u \approx g \cos \theta$$

$$= (4)^2 + (2.73 \times 10^{-3} \times 980)^2 = 16 + 7.16 = 23.16$$

$$u_o = 4.81 \text{ cm/sec}$$

Expressing the equation of motion in terms of τ and u_0 gives

$$\tau \frac{d\vec{v}}{dt} + \vec{v} = \vec{u}_o \tag{6.6}$$

Suppose the Cartesian coordinates are aligned such that at $t = 0$ the particle is at the origin. In addition, the coordinates are rotated so that the x-axis is parallel to $\vec{u}_0$. Finally, the initial velocity vector of the particle is oriented such that it lies in the x-y plane. Then this initial velocity vector can be broken down into x and y velocity components, $\vec{v}_{x_i}$ and $\vec{v}_{y_i}$, that is $\vec{v}_{x_i} + \vec{v}_{y_i} = \vec{v}_i$. Figure 6.2 illustrates the general orientation for solution of Eq. (6.6).

It should be realized that this coordinate system can be rotated at will. Although the orientation chosen is for convenience in solving the equation, it does

Figure 6.2. General orientation for solution of Equation (6.6). u_0 is aligned so it is parallel to X-axis.

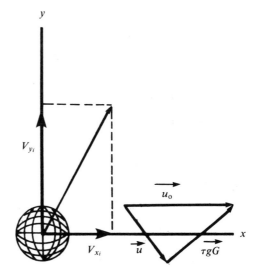

not necessarily reflect the actual physical orientation of the problem (gravity may not be down, for example). Hence in using this development, it is helpful to keep actual particle orientation in mind.

Equation (6.6) in its vector form represents two scalar differential equations, one representing motion in the x-direction and the other representing motion in the y-direction:

$$\tau \frac{dv_x}{dt} + v_x = u_o \tag{6.7a}$$

$$\tau \frac{dv_y}{dt} + v_y = 0 \tag{6.7b}$$

Integration of these equations with the initial conditions that $x = 0$ and $y = 0$, and $v_x = v_{x_i}$ and $v_y = v_{y_i}$ at $t = 0$, gives two equations for the velocity of the aerosol particle at any time,

$$v_x = u_o + (v_{x_i} - u_o)e^{-t/\tau} \tag{6.8a}$$

$$v_y = v_{y_i} e^{-t/\tau} \tag{6.8b}$$

and two equations for the particle's position

$$x = u_0 t + \tau(v_{x_i} - u_0)(1 - e^{-t/\tau}) \tag{6.9a}$$

$$y = v_{y_i} \tau(1 - e^{-t/\tau}). \tag{6.9b}$$

These four equations completely describe the position and trajectory of the particle at any time.

EXAMPLE 6.4

A 30-μm diameter unit density sphere ($\tau = 2.73 \times 10^{-3}$ seconds) falling at a terminal settling velocity of 2.7 cm/sec is captured by a horizontal air flow of 100 ft/min which is flowing into a hood. Find its velocity one millisecond later, relative to the point at which it was captured.

First it is necessary to rotate axes so that vector u_0 lies along x-axis. Then u_0 will have a value of

$$u_0 = \sqrt{(50.83)^2 + (2.7)^2} = 50.90 \text{ cm/sec}$$

The rotation required is $3°$. Then

$$v_{x_i} = 2.7 \sin 3° = 0.14$$

$$v_{y_i} = 2.7 \cos 3° = 2.70$$

$$e^{-t/\tau} = \exp(-10^{-3}/2.73 \times 10^{-3}) = 0.693$$

velocity:

$$v_x = u_o + (v_{x_i} - u_o) \exp(-t/\tau)$$

$$v_x = (50.90) + (0.14 - 50.90)(0.693)$$

$$v_x = 15.71 \text{ cm/sec}$$

$$v_y = v_{y_i} e^{-t/\tau}$$

$$v_y = (2.70)(0.693) = 1.87 \text{ cm/sec}$$

It is now necessary to switch back from the artificial coordinate system to the real one. This can be done as follows:

$$v_V = 15.71 \sin 3° + 1.87 \cos 3° = 2.70 \text{ cm/sec}$$

$$v_H = 15.71 \cos 3° - 1.87 \sin 3° = 15.59 \text{ cm/sec}$$

To get the particle position, a similar approach would be taken.

Notice that for typical aerosol particle sizes, the exponential terms rapidly disappear. Note also that the particle is rapidly acquiring the velocity of the horizontal airflow.

TERMINAL SETTLING VELOCITY

Equations 6.8 and 6.9 can now be applied to the case of a particle falling under the influence of gravity in still air ($u = 0$). Since the direction of the gravitational force is along the x-axis, Eq. (6.8a) shows that even with an initial velocity component in some other direction, eventually the only velocity the particle will have will be in the direction of the gravitational force. The velocity in the direction of gravity is given by

$$v_g = \tau g + (v_{x_i} - \tau g)e^{-t/\tau}$$

As time progresses the particle will attain a constant velocity given by τg, known as the terminal settling velocity of the particle. If a particle is initially given a velocity greater than this, it will decelerate until it has reached τg. If the particle's initial velocity is less, it will increase to a value of τg. If the particle falls from rest, it will accelerate until τg is attained. Thus τg is called the terminal settling velocity of a particle v_g,

$$v_g = \tau g \qquad (6.10)$$

EXAMPLE 6.5

An asbestos fiber is reported to have an aerodynamic diameter of 1.79 μm. Determine its terminal settling velocity.

$$\tau = \frac{1}{18} \frac{d_a^2}{\mu} = \frac{1(1.79 \times 10^{-4})^2}{[18]\, 1.83 \times 10^{-4}}$$

$$\tau = 9.73 \times 10^{-6} \text{ sec}$$

$$v_g = \tau g = 9.73 \times 10^{-6} \times 980$$

$$v_g = 9.53 \times 10^{-3} \text{ cm/sec}$$

If the Cunningham factor had not been ignored, the estimated settling velocity would have been about 35% greater. This error increases with decreasing size.

Sometimes particles are so small that is is necessary to consider Cunningham's correction factor. Then

$$v_g = \tau g C_c \tag{6.11}$$

EXAMPLE 6.6

Determine the terminal settling velocity of a 0.5-μm-diameter silica sphere (ρ = 2.65 g/cm^3). Include the Cunningham correction factor in the estimate.

From Eq. (5.3),

$$C_c = \left[1 + \frac{2\lambda}{d}(1.257)\right] \cong 1 + \frac{(2)(7 \times 10^{-6})}{0.5 \times 10^{-4}}(1.257)$$

$$C_c = 1.35$$

$$\tau = \frac{1}{18} \frac{d^2}{\mu} \rho_p = \frac{1}{18} \frac{(5 \times 10^{-5})^2 (2.65)}{1.83 \times 10^{-4}} = 2.01 \times 10^{-6} \text{ sec}$$

$$v_g = \tau g C_c = (2.01 \times 10^{-6})(980)(1.35) = 2.66 \times 10^{-3} \text{ cm/sec}$$

Equation (6.11) is often derived by merely equating Stokes' resistance with the gravitational force. Although conceptually simpler, it does not provide the insights into the time-dependent cases of acceleration or deceleration of the particle to terminal velocity. The rapidity with which the terminal settling velocity is reached is given the factor $e^{-t/\tau}$. Thus the smaller the value of τ, the more quickly an aerosol particle will reach equilibrium or steady-state conditions. For example, for a 2-μm-diameter unit density sphere τ has a value of 1.305×10^{-5} sec. Since e^{-7} is about 0.001, equilibrium values are essentially reached when $\tau = 7$, or for the 2-μm sphere, within about 100 μsec. Table 6.1 gives values of τ and 7τ for unit density spheres of other diameters. It is clear that particles smaller than several micrometers in diameter will rapidly accelerate or decelerate

Table 6.1. Relaxation times (τ) and equilibrium time (7τ) for unit density spheres at 1 atmosphere pressure and 20°C

Diameter, μm	Relaxation time, sec	Equilibrium time, sec
0.05	4.00×10^{-8}	2.80×10^{-7}
0.1	9.15×10^{-8}	6.41×10^{-7}
0.5	1.03×10^{-6}	7.19×10^{-6}
1.0	3.57×10^{-6}	2.50×10^{-5}
5.0	7.86×10^{-5}	5.50×10^{-4}
10.0	3.09×10^{-4}	2.16×10^{-3}
50.0	7.62×10^{-3}	5.33×10^{-2}

to equilibrium conditions, so that generally for these sizes of particles it is possible to neglect the inertial term in Eq. (6.1).

STOP DISTANCE

Consider the case of a particle having an initial velocity v_{y_i} in the y-direction when u_o is zero. This is equivalent to a particle being projected into still air, neglecting gravity. It can be seen from Eq. (6.8b) that the particle rapidly decelerates to zero velocity. While decelerating the particle traverses a distance which can be found from Eq. (6.9b) when t goes to infinity. This distance,

$$y = v_{y_i}\tau \qquad (6.12)$$

is known as the *stop distance* or *horizontal range* of the particle. Equation (6.12) indicates that small particles move very short distances before coming to rest; a 1-μm-diameter particle projected into air at an initial velocity of 1000 cm/sec, for example, moves a distance of only 0.0036 cm before stopping.

EXAMPLE 6.7

Determine the stop distance of a 2-μm-diameter unit density sphere which is projected into still air with an initial velocity of 1000 cm/sec. (Neglect gravity.)

$$\tau = \frac{1}{18} \frac{d^2}{\mu} \rho_p C_c$$

$$\tau = \frac{1}{18} \frac{(2 \times 10^{-4})^2}{1.83 \times 10^{-4}} (1) \left[1 + \frac{2 \times 7 \times 10^{-6}}{2 \times 10^{-4}} (1.257) \right]$$

$$\tau = (1.21 \times 10^{-5})(1.09) = 1.32 \times 10^{-5} \text{ sec}$$

$$y = (1000)(1.32 \times 10^{-5}) = 1.32 \times 10^{-2} \text{ cm}$$

PARTICLE ACCELERATION OR DECELERATION

For particles injected into a moving airstream (similar to acceleration under the influence of gravity and similar to problems of particle deceleration), it can be seen that the difference between particle velocity and stream velocity decreases by a factor of e for each time period $t = \tau$. Thus within 7τ steady-state conditions are reached.

EXAMPLE 6.8

A 40-μm-diameter unit density sphere falls across a slot opening for a ventilation system into which air is being drawn. How long will it take the particle to achieve the velocity of the inrushing air?

$$\tau = \frac{1}{18} \frac{d^2}{\mu} \rho_p = \frac{1}{18} \frac{(40 \times 10^{-4})^2}{1.83 \times 10^{-4}} (1) = 4.86 \times 10^{-3} \text{ sec}$$

$$t = 7(4.86 \times 10^{-3}) \text{ sec} = 0.034 \text{ sec}$$

This indicates that within 0.034 sec the particle will be caught up and transported by the moving airstream. Although this seems like a fairly short time, it may be too long to ensure capture of the particle. It should be apparent from this analysis that small particles ought to be easier to capture with a ventilation system than large particles.

LIMITATIONS

Equations of motion presented here were developed for cases of uniform medium velocity, and are oversimplified for many other cases regarding aerosols. In addition, evaluation of the equations for the trajectories of aerosol particles is sometimes impossible because of the difficulty in accurately describing the field of flow. Although for laminar flow Eq. (6.6) can be separated into x- and y-components, with increasing Reynolds number the nonlinearity of the resisting force prevents separation of the vector equation. Fortunately, most aerosol problems can be treated in the low Reynolds number regime.

EXAMPLE 6.9

Determine the diameter of a unit density sphere that has a Reynolds number at terminal settling velocity equal to one.

$$\text{Re} = \frac{v_g d}{0.152}$$

$$v_g = \tau g = \frac{1}{18} \frac{d^2}{\mu} \rho_p g$$

Substituting and rearranging gives

$$d^3 = \frac{(0.152)(18)(Re)(\mu)}{\rho_p g} = \frac{(0.152)(18)(1)(1.83 \times 10^{-4})}{(1)(980)}$$

$$d^3 = 5.11 \times 10^{-7}$$

$$d = 79.94 \times 10^{-4} \cong 80 \; \mu m$$

This represents a rough guide for the upper size of particles for which Stokes' law applies. This size will be different for particles of different densities.

ONE-DIMENSIONAL MOTION AT HIGH REYNOLDS NUMBERS

There are occasions when particle motion is so great or particle diameter so large that Stokes' law is no longer applicable. Then some other simplifying approach must be taken. In Chapter 4 this problem was treated for the case of sedimenting particles through the use of plots of $C_D Re^2$ versus Re and C_D/Re versus Re.

For the generalized case of one-dimensional particle motion, recall that

$$F_D = A C_D \rho_m \frac{v^2}{2} \qquad (6.13)$$

$$Re = \frac{v d \rho_m}{\mu} \qquad (6.14)$$

$$\frac{dv}{dRe} = \frac{\mu}{\rho_m d} \qquad (6.15)$$

From Eq. (6.1)

$$m \frac{dv}{dt} = F \qquad (6.16)$$

so, for spheres,

$$\frac{\pi}{6} d^3 \rho_p \frac{dv}{dt} = \frac{\pi}{4} d^2 C_D \rho_m \frac{v^2}{2}$$

$$\frac{dv}{dt} = \frac{3 C_D \rho_m v^2}{4 d \rho_p} \qquad (6.17)$$

Expressing Eq. (6.17) in terms of the Reynolds number

$$\frac{dRe}{dt} = \frac{3 C_D \mu}{4 \rho_p d^2} Re^2 \qquad (6.18)$$

gives, on inverting,

$$dt = \frac{4\rho_p d^2}{3\mu}\left[\frac{d\text{Re}}{C_D \text{Re}^2}\right] \tag{6.19}$$

On integration from Re_i to Re_f, Eq. (6.19) yields

$$t = \frac{4}{3}\frac{\rho_p d^2}{3\mu}\int_{\text{Re}_i}^{\text{Re}_f}\frac{d\text{Re}}{C_D \text{Re}^2} \tag{6.20}$$

Similarly, for displacement, since $s = vt$,

$$s = \frac{4}{3}\frac{\rho_p d^2}{\mu}\int\frac{d\text{Re}}{C_D \text{Re}^2}\left[\frac{\mu \text{Re}}{\rho_m d}\right] \tag{6.21}$$

$$s = \frac{4}{3}\frac{\rho_p}{\rho_m}d\int_{\text{Re}_i}^{\text{Re}_f}\frac{d\text{Re}}{C_D \text{Re}} \tag{6.22}$$

The utility of Eq. (6.20) and (6.22) lies in their generality. All that is required is an expression for C_D for the range of Reynolds numbers over which the particle is moving.

EXAMPLE 6.10

A 6-in.-diameter grinding wheel is operated at 1750 rpm. How far will a 0.1-mm particle be thrown from the wheel if gravity is neglected. Assume a spherical particle having a density of 3 g/cm^3.

Initial velocity $= \omega r = 1750 \times \dfrac{1}{60} \times 2\pi \times 6 \times 2.54 \times \dfrac{1}{2} = 1396$ cm/sec

$$\text{Re}_i = \frac{dv}{0.152} = \frac{(0.01)(1396)}{0.152} = 91.9$$

From Eq. (4.7)

$$C_D = 14/\text{Re}^{0.5}$$

This can only be used for $2 < \text{Re} < 800$.

$$s = \frac{4}{3}\frac{\rho_p}{\rho_m}d\int_{\text{Re}_i}^{\text{Re}_f}\text{Re}^{0.5}\frac{d\text{Re}}{14}\text{Re} = \frac{4}{3}\frac{1}{14}\frac{\rho_p}{\rho_m}d\int_{\text{Re}_i}^{\text{Re}_f}\text{Re}^{-1/2}\,d\text{Re}$$

Integrating gives

$$s = \frac{2}{21}\frac{3}{0.0012}(0.01)[\text{Re}^{1/2}]\,\Big|_{9.19}^{2} = 2.38[\sqrt{2} - \sqrt{91.9}]$$

$$= -19.46 \text{ cm}$$

To go from Re = 2 to Re = 0, use C_D = 24/Re

$$s = 2.38 \int_2^0 \frac{Re}{24} dRe = \frac{2.38}{24} \times [Re] \Big|_2^0$$

$$s = \frac{2.38}{24}(-2) = -0.20$$

Total stop distance = 19.46 + 0.20 = 19.66 cm

Notice that the particle travels the greatest distance at the higher Reynolds numbers. The laminar flow contribution to the stop distance is small compared to the intermediate or turbulent flow contribution.

IDEAL STIRRED SETTLING

Although with an aerosol each particle will settle at its own terminal settling velocity, settling rarely takes place in absolutely still air since there is always some circulation and mixing. This mixing has the effect of producing a uniform aerosol concentration which decreases with time because of sedimentation.

Consider a cylinder of uniform cross-sectional area A and height H, filled with n_o particles of a monodisperse aerosol. With mixing, but in the absence of any other external forces, the number of particles moving in an upwards direction and crossing a plane L which cuts the cylinder at some arbitrary height will equal the number moving downward. That is,

$$dn = 1/2 \, CAv_u \, dt - 1/2 \, CAv_d \, dt = 0 \qquad (6.23)$$

where C is the number of particles per unit volume and v_u and v_d the particle velocities up and down respectively.

Now suppose a negative force field is included (i.e., include the force due to gravity which imparts an additional downward velocity v_g on the particles; this is the terminal settling velocity). Then

$$dn = 1/2 \, CA(v_u - v_g)dt - 1/2 \, CA(v_d + v_g)dt \qquad (6.24)$$

since $v_u = v_d$

$$dn = -v_g CA \, dt \qquad (6.25)$$

Also

$$n = CAH, \qquad dn = AHdC$$

so that

$$\frac{dC}{c} = -\frac{v_g}{H} dt \qquad (6.26)$$

Assuming v_g is independent of time, position, and concentration, Eq. (6.26) can be integrated to give

$$C = C_0 \exp \frac{-v_g t}{H} \qquad (6.27)$$

where C_0 is the aerosol concentration at $t = 0$. Equation (6.27) implies exponential decay of an aerosol concentration in a closed chamber. This is observed in practice.

EXAMPLE 6.11

The smoke concentration in a room is found to be 50 mg/m^3. If this aerosol is made up of 0.75-μm-diameter spherical particles (unit density), estimate the concentration in the room three hours later. The ceiling height is 8 ft.

For 0.75-μm spheres,

$$v_g = \frac{1}{18} \frac{d^2}{\mu} \rho_p C_c g$$

$$= \frac{1}{18} \frac{(0.75 \times 10^{-4})^2}{1.83 \times 10^{-4}} (1) \left[1 + \frac{0.14}{0.75} (1.257) \right] 980$$

$$v_g = 2.07 \times 10^{-3} \text{ cm/sec}$$

$$H = 8 \text{ ft} \times 30.5 \text{ cm/ft} = 244 \text{ cm}$$

$$C = C_0 \exp(-v_g t/H) = 50 \exp\left(-\frac{2.07 \times 10^{-3}}{244} \, 3 \times 60 \times 60 \right)$$

$$C = 50 \times 0.912 = 45.6 \text{ mg/m}^3$$

For polydisperse aerosols, concentration decay can be estimated by treating each size independently. A graph of room concentration versus time can be made from a composite of individual exponential decay curves.

PROBLEMS

1. Show that if $v_y = v_{y_i}$ at $t = 0$, the solution to

$$\tau \frac{dy}{dt} + v_y = 0$$

is

$$v_y = v_{y_i} e^{-t/\tau}$$

Then using the relationship $v_y = dy/dt$ show that

$$y = v_{y_i} \tau (1 - e^{-t/\tau})$$

if

$$y = 0 \text{ at } t = 0$$

2. Compute the value of τ for a 15-μm-diameter sand particle ($\rho = 2.65$ g/cm^3) Then compute:
   a. its terminal settling velocity
   b. its Reynolds number at this velocity
   c. its stop distance

3. a. What is the diameter and terminal settling velocity of a unit density sphere having Re = 0.5 at its terminal settling velocity?
   b. Assuming the particle initially started from rest, how long will it take to reach 1/2 its terminal settling velocity?

4. A 10-μm gold sphere ($\rho = 19.3$ g/cm^3) is dropped from a 10-ft-high platform. Estimate the time it takes to strike the ground:
   a. neglecting air resistance
   b. including air resistance

5. A 200-μm-diameter raindrop falls freely in the atmosphere. Determine its terminal settling velocity. Compare this to the measured value given in Table 5.2.

6. What is the diameter of a unit density particle which falls with a terminal settling velocity of 200 cm/sec.

7. Using Stokes' law show that

$$C_D = \frac{24}{Re}$$

8. An approximate formula sometimes used to estimate the settling velocity in feet per minute of airborne dust particles is

$$v(\text{fpm}) = \frac{d^2}{100}$$

where d is the particle diameter in μm. Compare the estimate given by this equation to the terminal velocity given by Stokes' law for 10-, 1.0- and 0.1-μm particles with a density of 2.3.

9. Given three unit density spherical particles of 2, 0.2, and 0.02 μm diameter, compute the sedimentation velocity for each
   a. neglecting C_c
   b. correcting for C_c

10. Examine the settling velocity of a 100-μm unit density sphere with and without the correction

$$\left(1 + \frac{3}{16} Re\right)$$

applied. How much error is introduced by neglecting this correction?

11. Determine the position of the particle described in Example 6.4 1 msec. after it is captured.

SEVEN

PARTICLE KINETICS
Impaction, Centrifugation,
Respirable Sampling,
Isokinetic Sampling

CURVILINEAR MOTION

When particles are transported by air currents, changes in the direction of these currents give rise to accelerating forces on the aerosol particles. Thus spinning of air tends to move the aerosol away from the axis of rotation or rapid changes of airflow around an obstacle can result in aerosol particles being deposited on that body. This may be one of the principal mechanisms by which particles are removed by nature from the atmosphere. Sampling and collection devices such as impactors or impingers are based on the use of centrifugal forces as are such other devices as "cyclones" and aerosol centrifuges.

The magnitude of the accelerating force that acts on a particle in curvilinear motion depends on particle inertia. The greater the inertia of the particle, the greater will be the displacement. Inertia depends on particle mass and velocity. Heavy particles will be displaced more from the streamlines in which they are traveling than light ones, and increases in velocity will increase displacement for a particle of given mass.

If a particle is moving in a circular path around a point a distance r away with an angular velocity of ω, it experiences a radial acceleration of

$$a_r = \omega^2 r \tag{7.1}$$

and a tangential velocity of

$$v_\omega = \omega r \tag{7.2}$$

If the particle is moving in air (at low Re), its motion will be resisted by a force equal to the Stokes' resistance. Equating the radial accelerating force with the resisting force gives

$$F_r = F_R \tag{7.3}$$

$$ma_r = 3\pi\mu d v_r / C_c$$

solving for v_r assuming a spherical particle yields,

$$v_r = \frac{\dfrac{\pi}{6} d^3 \rho_p \omega^2 r C_c}{3\pi\mu d} = \tau\omega^2 r \tag{7.4}$$

$$v_r = \frac{v_\omega^2}{r} \tau \tag{7.5}$$

resulting in an expression for the radial velocity of a particle.

EXAMPLE 7.1

A 10-μm-diameter unit density sphere is held in a circular orbit by an electric field. The orbit is 25 cm in radius and the particle moves around the center of the circle at a rate of 100 rpm.

a. What is the radial velocity of the particle at the instant the electric field is removed?

From Eq. (7-4)

$$v_r = \tau\omega^2 r = \frac{d^2}{18\mu}\rho C_c \omega^2 r$$

$$= \frac{1}{18}\frac{(10 \times 10^{-4})^2}{1.83 \times 10^{-4}}(1)(1)\left(\frac{100}{60} \times 2\pi\right)^2(25)$$

$$= 0.832 \text{ cm/sec}$$

b. How far will the particle move until its radial velocity is dissipated?

The distance the particle will move is just the stop distance,

$$s = \tau v_r = (3.04 \times 10^{-4})(0.832)$$

$$v = 2.53 \times 10^{-4} \text{ cm}$$

IMPACTION OF PARTICLES

When air carrying particles suddenly changes direction, the particles, because of their inertia, tend to continue along their original paths. If the change in air direction is caused by an object placed in the airstream, particles with sufficient inertia will strike the object. This process is known as *impaction*. It is the mechanism by which many large particles are removed from the atmosphere, it is one of the important mechanisms for removal of particles by the lungs, and it is important in air cleaning as well as aerosol sampling. The process of impaction can be modeled using the equation of motion of an aerosol particle, if an appropriate choice is made of the velocity field. Often compromises have to be made in this choice. In any theoretical development, however, certain factors seem to be important.

Consider a simple model of impaction. Air issues from a long slot of width $2W$ at a velocity of u. A surface A-A is placed normal to the discharging flow a distance S away. With this configuration, air leaving the slot must make a $90°$ turn before it escapes. Particles that fail to make this turn strike or "impact" on surface A-A and are assumed to be retained by that surface.

As a crude first approach (Fuchs, 1964), it can be assumed that the streamlines of the air issuing from the slot are quarter circles with their centers at C (see Fig. 7.1) and that $S = W$. At point B a particle has the following velocities: a tangential velocity given by $v_\omega = u$ and a radial velocity given by

$$v_r = \frac{dr}{dt} = \frac{v_\omega^2}{r}\tau \tag{7.6}$$

In a time dt the particle will be displaced toward plane A-A at a distance

$$ds = \frac{u^2}{r}\tau \sin\phi\, dt \tag{7.7}$$

where ϕ is the angle formed by the line connecting points B and C and the plane normal to the airflow in the slot which passes through point C. As the streamlines turn from the slot to be parallel with the surface, ϕ goes from $0°$ to $90°$. The angle ϕ changes as

$$d\phi = \frac{v_\omega}{r} dt. \tag{7.8}$$

In traversing the full $90°$ curve the particle will be displaced a distance δ,

$$\delta = \int_0^{\pi/2} \omega\tau \sin\phi\, d\phi = u\tau, \tag{7.9}$$

that is to say, the particle will move one stop distance out of its streamline while losing all of its original velocity parallel to the slot.

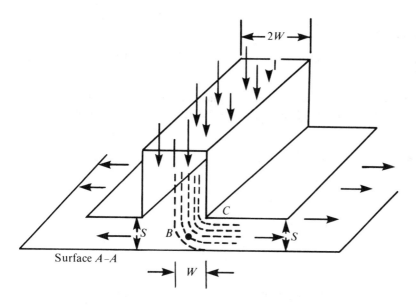

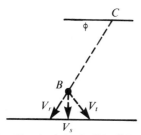

Figure 7.1. Sketch of simple "ideal" impactor.

Since all particles that lie within a distance δ of the slot centerline are considered to be removed, the overall removal efficiency, ϵ, of the impactor will be

$$\epsilon = \frac{\delta}{W} = \frac{u\tau}{S} \tag{7.10}$$

This is, of course, only a very crude approximation for impactor efficiency, since the actual flow field is much more complex, varying in configuration depending on slot Reynolds number. In general, $S \neq W$. This sample model does give some idea of how effective a particular impactor configuration can be, the estimate being reasonably good when $\epsilon \gtrsim 1$, but rapidly losing accuracy for $\epsilon < 0.7$.

EXAMPLE 7.2

A rectangular slot impactor has slot dimensions of 2.08-cm length and 0.358 cm width.

Estimate the flow in liters per minute required for this impactor in order that 15-μm-diameter unit density spheres can be collected with near 100% efficiency.

$$\tau = \frac{1}{18} \frac{d^2}{\mu} \rho_p C_c$$

$$= \frac{1}{18} \frac{(15 \times 10^{-4})}{1.83 \times 10^{-4}} (1)(1)$$

$$\tau = 6.83 \times 10^{-4} \text{ sec}$$

$$\epsilon = \frac{u\tau}{W}$$

$$u = \frac{\epsilon W}{\tau} = \frac{(1)\left(\dfrac{0.358}{2}\right)}{6.83 \times 10^{-4}}$$

$$u = 262 \text{ cm/sec}$$

$$Q = A \times u = (2.08)(0.358)(262) = 195 \text{ cm}^3/\text{sec}$$

$$Q = 11.71 \text{ lpm}$$

The quantity $u\tau/W$ is an important dimensionless parameter in impactor studies, known as the Stokes' number (Stk),

$$\text{Stk} = \frac{u\tau}{W} \qquad (7.11)$$

This dimensionless parameter is used to describe impactor behavior. For rectangular impactors, W is the impactor half-width; for circular openings, W represents the radius of the impactor opening. Some authors prefer to use the impaction parameter, Ψ, rather than the Stokes' number, to describe impactor properties (e.g. Green and Lane, 1964; Ranz and Wong, 1952). The impaction parameter is defined as

$$\Psi = \frac{u\tau}{2W} \qquad (7.12)$$

a factor which is one-half the Stokes' number. Thus in computing the impaction parameter, the slit width or jet diameter is used for the denominator.

EXAMPLE 7.3

Compute Stk, $\sqrt{\text{Stk}}$, ψ, and $\sqrt{\psi}$ for a circular jet impactor when $\tau u = 0.004$ cm, and the jet diameter is 0.1 mm.

$$\text{Stk} = \frac{\tau u}{W} = \frac{4 \times 10^{-3}}{0.005} = 0.8$$

$$\sqrt{\text{Stk}} = 0.89$$

$$\psi = \frac{\text{Stk}}{2} = 0.4$$

$$\sqrt{\psi} = 0.63$$

It is common practice to plot impactor efficiency as a function of either $\sqrt{\text{Stk}}$ or $\sqrt{\psi}$ (e.g. Rao and Whitby, 1978). This is done because particle diameter is present in either term as d^2, making the square root of the term proportional to particle diameter.

Precise analysis of the flow field in impaction reveals that there exists a Stokes number below which impaction will not occur. For rectangular openings this value, $\text{Stk}_{cr} = 1/4$, and for circular jets, $\text{Stk}_{cr} = 1/8$ (Levin, 1963).

Impactor Operation

The characteristic behavior of impactors depends on factors such as nozzle to plate distance, nozzle shape, flow direction, and Reynolds numbers for both the jet and particle. Other factors of importance include the probability the particles will stick to the impaction surface and particle loss to the walls of the impactor. It is not surprising that with such a variety of possible variables it is quite difficult, if not impossible, to accurately predict impactor characteristics on purely theoretical grounds. For a round jet the Reynolds number calculation is straightforward: $\text{Re} = av/v$ where a is the jet diameter, v the velocity in the jet, and v the kinematic viscosity. For a flow Q cm^3/sec,

$$\text{Re} = \frac{4Q}{\pi v a} \qquad (7.13)$$

For a rectangular jet the "wetted perimeter" concept must be used (Marple, 1970). That is, the opening to be used in computing the Reynolds number is defined as $d = 4(\text{area})/\text{perimeter}$. For a rectangle of length L and width $2W$, $d = 4WL/(2W + L)$. When $L \gg W$,

$$\text{Re} = \frac{2Q}{Lv} \qquad (7.14)$$

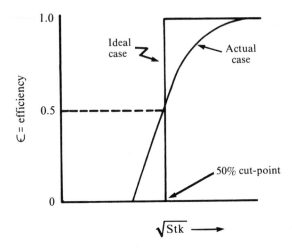

Figure 7.2. Typical impactor stage efficiency curve.

For a well-designed impactor, a typical plot of impactor efficiency versus $\sqrt{\text{Stk}}$ is shown in Fig. 7.2. It can be seen that the efficiency curve may deviate from the ideal case. In the ideal case, for all efficiencies there would be a single value of $\sqrt{\text{Stk}}$, and hence a sharp size cut of the impactor. All particles larger than this size would be collected and all smaller sizes would be passed. In actuality, this is not the case and there is a range of particle sizes collected with varying efficiencies. To represent the impactor stage collection characteristic, it is often the practice to choose the 50% efficiency point as the representative cut point. The maximum slope at this point most nearly represents the ideal case. In Fig. 7.2 both the actual and ideal cases would be considered to have the same characteristic cut size.

Figure 7.3 shows theoretical impactor performance when a number of parameters are varied, including jet-to-plate spacing, jet Reynolds number, and throat length to width ratio. These curves indicate that impactor efficiency is fairly insensitive to Reynolds number in the range $500 < \text{Re} < 25{,}000$ and that impactor efficiency is also relatively independent of S/W and T/W ratios, except for small values of S/W.

EXAMPLE 7.4

A round-jet impactor is operated such that the jet Reynolds number is 3000. Using Fig. 7.3b, find the particle diameter (unit density sphere) that will be collected with 50% efficiency if the jet diameter is 0.3 cm.

$$\text{Re} = \frac{av}{0.152} = \frac{(0.3)(v)}{0.152} = 3000$$

$$v = 1520 \text{ cm/sec}$$

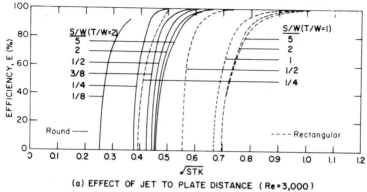

(a) EFFECT OF JET TO PLATE DISTANCE (Re = 3,000)

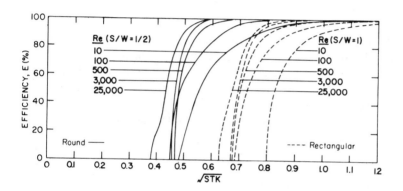

(b) EFFECT OF JET REYNOLDS NUMBER (T/W = 1)

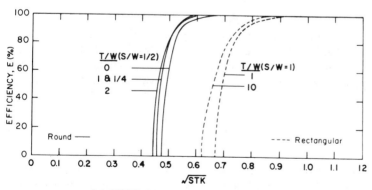

(c) EFFECT OF THROAT LENGTH (Re = 3,000)

Figure 7.3. Impactor efficiency curves for rectangular and round impactors showing effects of jet-to-plate distances in Reynolds number, Re, and throat length T. W is impactor half-width or impactor jet radius. (From Marple and Willeke, 1979.) (a) Effect of jet to plate distance (Re = 3000). (b) Effect of jet Reynolds number ($T/W = 1$). (c) Effect of throat length (Re = 3000).

From Fig. 7.3b, at Eff = 50%, $\sqrt{\text{Stk}} = 0.46$

$$\text{Stk} = 0.21 = \frac{\tau v}{0.3/2}$$

$$\tau = \frac{(0.21)(0.15)}{1520} = 2.07 \times 10^{-5} \text{ sec}$$

$$\tau = \frac{1}{18} \frac{d^2}{\mu} (\rho) \text{ (neglect } C_c)$$

$$d^2 = \frac{18\,\tau\mu}{\rho} = \frac{(18)(2.07 \times 10^{-5})(1.83 \times 10^{-4})}{(1)}$$

$$d = 2.61 \times 10^{-4} \text{ cm} = 2.61 \text{ } \mu\text{m}$$

Theoretical impactor performance data can also be expressed in terms of the 50% cut size. Figure 7.4a shows $\sqrt{\text{Stk}_{50\%}}$ plotted as a function of S/W ratio and Fig. 7.4b shows the same ordinate plotted as a function of Re. These curves again illustrate the relative insensitivity of the factors S/W and Re on impactor performance, except in the extremes.

Impactor 50% cut points can be estimated from the equations

$$d_{50} = \sqrt{\text{Stk}} \sqrt{\frac{36\mu L}{C_c \rho_p Q} W} \tag{7.15}$$

for rectangular jets of length L and width $2W$, or

$$d_{50} = \sqrt{\text{Stk}} \sqrt{\frac{9\pi\mu a^3}{4C_c \rho_p Q}} \tag{7.16}$$

for round jets of diameter a. These are rearrangements of Eq. (7.11).

As shown in Fig. 7.4a, theoretical estimates of $\sqrt{\text{Stk}_{50}}$ are 0.71 for rectangular jet impactors and 0.46 for round jet impactors. These estimates represent the case of no particle reentrainment. With reentrainment, values of $\sqrt{\text{Stk}}$ will be higher.

EXAMPLE 7.5

In a calibration experiment 1.8-μm-diameter polystyrene spheres ($\rho = 1.05$ g/cm^3) were removed by a round jet impactor ($a = 0.1$ cm) from an airstream with 50% efficiency when the jet velocity was 2000 cm/sec. When a thin coating of a silicon grease was applied to the impactor surface, 50% collection was achieved at a velocity of 1600 cm/sec.

Compute the value of $\sqrt{\text{Stk}}$ for these two cases.

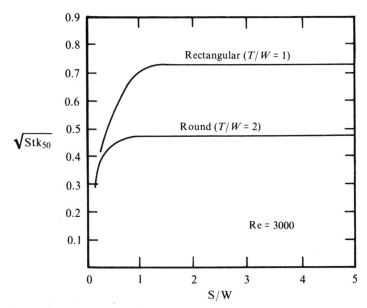

Figure 7.4a. 50% cut-off Stokes number as a function of jet-to-plate distance. (From Marple, 1970.)

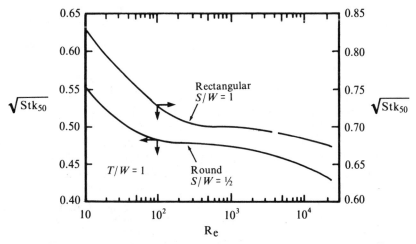

Figure 7.4b. 50% cut-off Stokes number as a function of Reynolds number. (From Marple, 1970.)

Compute using Eq. (7.16).

$$d_{50} = \sqrt{Stk_{50}} \sqrt{\frac{9\pi\mu a^3}{4C_c\rho_p Q}}$$

$$d_{50} = \sqrt{Stk_{50}} \sqrt{\frac{9\mu a}{C_c\rho_p v}}$$

$$C_c = 1 + \frac{2\lambda}{d}(1.257) = 1 + \frac{0.14}{1.8}(1.257) = 1.10$$

$$\sqrt{Stk_{50}} = d_{50} \sqrt{\frac{C_c\rho_p v}{9\mu a}}$$

For a clean surface

$$\sqrt{Stk_{50}} = 1.8 \times 10^{-4} \sqrt{\frac{(1.10)(1.05)(2000)}{9(1.83 \times 10^{-4})(0.1)}}$$

$$\sqrt{Stk_{50}} = 0.67$$

For greased surface

$$\sqrt{Stk_{50}} = 1.8 \times 10^{-4} \sqrt{\frac{(1.10)(1.05)(1600)}{9(1.83 \times 10^{-4})(0.1)}}$$

$$\sqrt{Stk_{50}} = 0.60$$

This result requires some thought. The lower value of $\sqrt{Stk_{50}}$ for the greased slide indicates that for a specific set of impactor operating parameters, using an adhesive layer on the impactor plate has the effect of *lowering* the 50% cut point.

Internal deposition of material may take place within the impactor and not on the impactor stage. Consider the following experiment:

An impactor is operated such that there is no rebound and internal deposition can be halted by the presence of an electric field. If, when the impactor is operated with the field on, a certain slot velocity gives a downstream concentration of a monodisperse aerosol that is 50% of the upstream concentration, what happens when the field is shut off so wall deposition can take place?

Since deposition is occurring, the downstream aerosol concentration will drop. In order to increase this concentration so that C/C_o will again be 50%, the impactor flow must be reduced. Thus wall deposition has the effect of raising the value of $\sqrt{Stk_{50}}$ compared to the case of no wall deposition.

Analysis of Impactor Data

The most common configuration for impactors used for aerosol sampling is to have a series of jets of decreasing size, arranged so that the air passes in series from the largest through the smallest slot. This cascade arrangement permits the aerosol to be fractionated into a number of size intervals, depending on the number of impactor stages used. Aerosol mass collected on the different impactor stages is then analyzed to provide size distribution information.

Particle distribution data can be presented as a bar chart where the mass of material collected in the ith stage, m_i, is taken as the mass of particles in the size range $(d_{50})_{i+1}$ to $(d_{50})_i$. The height of the bar then represents the percentage of mass in that interval, and the size range gives the width of the bar.

EXAMPLE 7.6

A six-stage impactor yields the following data (a filter is placed downstream of the impactor as a final stage to collect all particles which might otherwise escape):

Stage No.	d_{50}, μm	Particle Mass Collected, μg	% in Interval
1	18	0	0
2	11	15	3.30
3	4.4	35	7.69
4	2.65	110	24.18
5	1.7	190	41.76
6	0.95	80	17.58
Filter	—	25	5.49

Plot a histogram showing the particle size distribution.

Computing the % in each interval, and assuming that the lower collection limit of the filter is 0 μm; Fig. 7.5 can be plotted. A second method is to plot the data in a form similar to Example 2.3.

In many cases a mean (or median) particle size and a standard deviation are desired.

The most common method of data reduction is to plot the stage data on log-probability paper. If a straight line can be fitted to the data a log-normal distribution is assumed. For this plot, the 50% cut diameter or median size for a given stage is taken as the characteristic size for that stage. That is, all particles equal to or larger than d_{50_i} are retained on the ith stage, all smaller particles pass through the stage. Cumulative percentage less than a given size is plotted on log probability paper as a function of particle size, and a line of best fit is drawn through the points. From this line a median diameter and geometric standard deviation can be determined. If particle mass measurements are used as estimates of material deposited on the various stages, and the cut diameters are in terms of aerodynamic diameter, the resultant median value is the *mass median aerodynamic diameter,* sometimes abbreviated as **MMAD.**

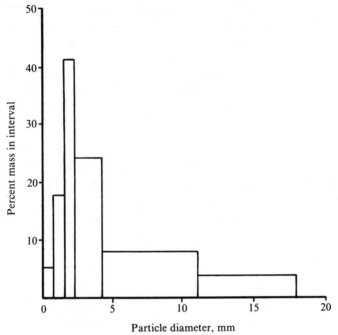

Figure 7.5. Histogram of impactor data.

EXAMPLE 7.7

Given the following impactor data, using a log-probability plot, determine MMAD and σ_g.

Stage No.	Aerodynamic d_{50i}	Mass, μg, Collected On Stage
0	20	0
1	10	0
2	5	10
3	3	35
4	2	70
5	1	190
6	0.8	60
7	0.5	85
Filter	0	50

Compute the cumulative percentage collected for each stage.

Interval	% in Interval	Cumulative % Less than Upper Interval Size
0- 0.5	10	10
0.5- 0.8	17	27
0.8- 1.0	12	39
1.0- 2.0	38	77
2.0- 3.0	14	91
3.0- 5.0	7	98
5.0-10.0	2	100

From the plot (Fig. 7.6) an MMAD of 1.2 μm is found and a σ_g of 2.0 determined. This is a mass median diameter because the weight or mass of aerosol collected on each stage was used in the analysis.

Figure 7.6.   Log-normal plot of impactor data given in Problem 7.7.

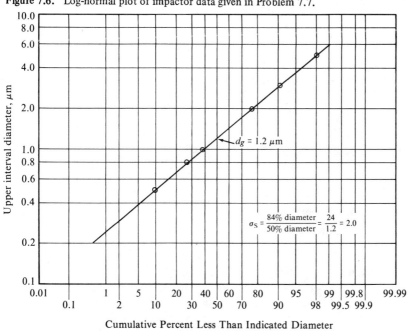

Errors Associated with Impactor Data

As mentioned earlier, particles can be deposited within the impactor housing or be reentrained from an impactor stage after collection. Both of these phenomena give rise to errors in impactor measurements. Placing adhesive on the impactor stages improves collection, as does the use of an impact adsorbing substrate for the impaction surface. Unwanted collection within the impactor can only be improved by impactor design.

Impactor calibrations must be done carefully to minimize error. In many cases this is not done and one should be wary of unsubstantiated claims of impactor performance.

Finally, the methods of data analysis presented here are quite crude. Fitting a straight line to data on log probability paper requires the assumption that the data are log-normally distributed, which may not be the case. However, more detailed and sophisticated methods such as the use of iterative calculational schemes seem unwarranted until more accurate calibration and measurement data are available (Fuchs, 1978).

CENTRIFUGATION OF PARTICLES

As seen earlier, terminal settling velocities of aerosol particles are generally quite small, being essentially negligible for particles less than 1 μm or so in diameter. One way to remove these particles from air is to subject them to high centrifugal forces. If an aerosol particle is caused to move in a circular path, it will have a radial acceleration given by Eq. (7.1). This radial accleration can be likened to the acceleration due to gravity in a gravitational field. By rotating the aerosol, accelerations many times that of gravity can be achieved.

EXAMPLE 7.8

A centrifuge has a radius of 50 cm and is operated at 500 rpm. Determine the ratio of radial acceleration to gravitational acceleration in this case.

$$\frac{\text{Radial acceleration}}{\text{Gravitational acceleration}} = \frac{\left(\frac{500}{60} \times 2\pi\right)^2 (50)}{980} = 1.40 \times 10^2$$

As can be seen from Example 7.8, quite large radial accelerations are possible, indicating that very small particles can be removed in this manner.

In an aerosol centrifuge particles are made to follow circular paths until they strike the outer wall of the unit. The distance from the inlet that a particle is deposited is indicative of particle size. This can be shown as follows:

Consider a centrifuge having an inner radius R_1 and an outer radius R_2. A particle enters at R_1 and travels across the annulus to be deposited somewhere on the surface at R_2. The radial velocity of the particle is given by Eq. (7.4)

$$v_r = \tau \omega^2 r \qquad (7.17)$$

that is,

$$\frac{dr}{dt} = \tau \omega^2 r \qquad (7.18)$$

The tangential velocity of the particle will be that of the airstream moving through the centrifuge. With a flow of Q cm^3/sec and a centrifuge channel depth of h

$$\bar{u} = \frac{Q}{A} = \frac{Q}{h(R_2 - R_1)} \qquad (7.19)$$

This represents an average velocity. In actuality the velocity distribution in the channel at any point r across the radius will be parabolic in shape, of a form given by

$$u = \frac{4k\bar{u}}{(R_2 - R_1)^2}(r + R_1)(R_2 - r) \qquad (7.20)$$

The factor k can range in value from 1.5 for deep, narrow channels to about 3 for rectangular ones (Tillery, 1979).

If l_D is the tangential distance the particle travels downstream before being deposited, then

$$dl_D = u\,dt \qquad (7.21)$$

In terms of r this becomes

$$dl_D = \frac{4k\bar{u}}{(R_2 - R_1)^2}\frac{1}{\omega^2 \tau r}(r - R_1)(R_2 - r)\,dr \qquad (7.22)$$

Integrating Eq. (7.22) between the limits R_1 and R_2 gives

$$l_D = \frac{2k\bar{u}}{\omega^2 \tau}\left[\frac{R_2 + R_1}{R_2 - R_1} + \frac{2R_1 R_2}{(R_2 - R_1)^2}\,2\ln\frac{R_1}{R_2}\right] \qquad (7.23)$$

EXAMPLE 7.9

Using Eq. (7.23), find the diameter of unit density spheres that would be deposited 2 cm from the entrance of a centrifuge that is operated at 300 rpm, with an average channel velocity of 100 cm/sec. The centrifuge inner diameter is 15 cm and the channel width is 1 cm. Assume $k = 2$.

$$l_D = \frac{2k\bar{u}}{\omega^2\tau}\left[\frac{R_2 + R_1}{R_2 - R_1} + \frac{2R_1R_2}{(R_2 - R_1)^2}\ln\frac{R_1}{R_2}\right]$$

$$2 = \frac{2[2][100]}{\left(\dfrac{300}{60} \times 2\pi\right)^2 \tau}\left[\frac{16 + 15}{16 - 15} + \frac{2(16)(15)}{(16 - 15)^2}\ln\frac{15}{16}\right]$$

$$\tau = 4.36 \times 10^{-3} \text{ sec} = \frac{1}{18}\frac{d^2}{\mu}\rho_p$$

$$d^2 = (4.36 \times 10^{-3})(18)(1.83 \times 10^{-4}) = 1.44 \times 10^{-5}$$

$$d = 37.9 \text{ } \mu m$$

Aerosol centrifuges are useful in the laboratory but have little application in air cleaning. When care is taken so that the aerosol enters the rotating annulus at a single point, the units are often called *aerosol spectrometers*.

CYCLONES

An accurate theory to predict cyclone behavior has yet to be achieved. In a cyclone, particle-laden air is introduced radially into the upper portion of a cylinder so that it makes several revolutions inside the cylinder before leaving axially along the cylinder centerline. While making these revolutions, the particles in the air are accelerated outward to the cylinder walls where they either stick and are retained (low particle loading) or are swirled down to a collection port at the bottom of the cylinder (high particle loading).

Important parameters in cyclone operation can be established by considering simple cyclone theory. Figure (7.7) shows a sketch of a typical cyclone. Air at a flow of Q cm^3/sec enters tangentially, revolving N_T times in the cyclone before it is discharged. Dust that is removed from the air spirals down into the dust discharge port.

Assuming that the gas moves through the cyclone as a rigid airstream with a spiral velocity equal to the average velocity at the cyclone inlet, the retention time for an element of gas within the cyclone is

$$t_r = N_t (2\pi R)/v_c \tag{7.24}$$

where v_c is the inlet velocity. During this retention time a particle can move a distance x across the width of the airstream B_c. Since the particle radial velocity is

$$v_r = \frac{v_c^2}{R}\tau \tag{7.25}$$

the time to go a distance x is

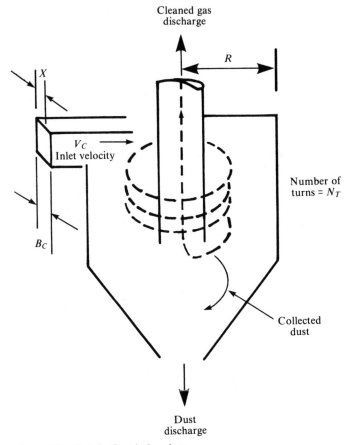

Figure 7.7. Sketch of typical cyclone.

$$t_x = \frac{xR}{v_c^2 \tau} \qquad (7.26)$$

and the time to go a distance B_c is

$$t = \frac{B_c R}{v_c^2 \tau} \qquad (7.27)$$

Equating this time with the transit time through the cyclone gives

$$\frac{B_c R}{v_c^2 \tau} = \frac{N_t(2\pi R)}{v_c}$$

$$\tau_{\text{crit}} = \frac{B_c}{v_c N_t 2\pi} \qquad (7.28)$$

This is the τ of the smallest particles that simple theory says should be collected with 100% efficiency. In terms of particle diameter

$$d_{crit} = \sqrt{\frac{9B_c\mu}{\rho_p v_c N_t \pi}} \qquad (7.29)$$

Cyclone efficiency, ϵ, is estimated from the ratio x/B_c

$$\epsilon = \pi N_t \frac{v_c \tau}{(B_c/2)} \qquad (7.30)$$

Since $v_c\tau/(B_c/2) =$ Stk, the Stokes' number for the cyclone, efficiency can be expressed as

$$\epsilon = \tau N_t (\text{Stk}) \qquad (7.31)$$

The factor N_T varies from 0.5 to about 10, depending on cyclone shape and size.

EXAMPLE 7.10

Estimate the collection efficiency for 5-μm unit density spheres in a small cyclone having a square entrance of 0.3 cm on a side when operated at a flow rate of 1.7 1pm. Use $N_T = 1$.

$$B_c/2 = \frac{0.3}{2} = 0.15 \text{ cm}$$

$$\tau = \frac{1}{18} \frac{d^2 \rho_p}{\mu} = \frac{1}{18} \frac{(5 \times 10^{-4})^2}{1.83 \times 10^{-4}} = 7.59 \times 10^{-5} \text{ sec}$$

$$v_c = \frac{Q}{A} = \frac{\dfrac{1.7 \times 1000}{60}}{0.3 \times 0.3} = 314.8 \text{ cm/sec}$$

$$\text{Stk} = \frac{v_c \tau}{B_c/2} = 0.159$$

$$\text{Eff} = \pi N_T \text{Stk} = 0.500 = 50\%$$

The efficiency predicted by Eq. (7.31) is only a rough estimate; the equation estimates a shape in the efficiency versus particle size curve that is different from what is actually observed. There are a number of factors not considered in this elementary derivation. First, laminar flow is assumed, but turbulent flow is often observed in practice. The effect of turbulence will be to move particles away from the cyclone walls or resuspend deposited ones. Hence turbulence will decrease cyclone efficiency. Second, the width of the cyclone inlet is not as impor-

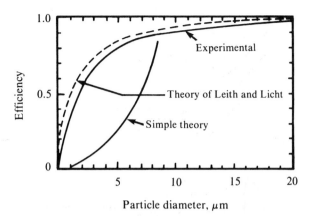

Figure 7.8. Comparison of simple theory, detailed theoretical efficiency, and experimental data for cyclone. (Leith and Licht, 1972.)

tant a parameter as overall cyclone diameter, since it is the width of an element of gas within the cyclone that determines particle deposition, and this width is not strongly controlled by inlet width. Finally, overall cyclone configuration will affect efficiency. This is not taken into account in the simple theory. Equation (7.31) does illustrate the general approach which has been followed in refining cyclone theory and, similar to the impactor and centrifuge equations, it permits rough estimates of system performance to be made. As can be seen in Fig. 7.8, however, the shape of the efficiency curve as predicted by simple theory is not consistent with experimental observations, the former being concave upwards and the latter concave downwards. Despite the apparent good fit of some theories to some experimental data (e.g. Leith and Licht, 1972), theories that are developed for large industrial-type cyclones do not give good predictions when applied to small, personal sampling-type cyclones. Chan and Lippmann, 1977, present a recent summary of current cyclone collection theories. Detailed criteria for cyclone design have recently been summarized by Licht (1980) and include many of the considerations mentioned above.

ISOKINETIC SAMPLING

In aerosol sampling the measured concentration and size distribution should represent as closely as possible the concentration and size distribution of the original aerosol. There are several reasons why the measured concentration can differ from the true concentration. One is that gravitational or inertial deposition of the sample as it flows into the sampling tube can result in loss of larger sized particles. Also, deposition or selective collection at the mouth of the sampling tube can yield either greater or lesser amounts of larger particles.

Consider the three cases shown in Fig. 7.9. In case (a) the probe is not aligned with flow. Because of inertia some particles may be lost by impaction, giving a

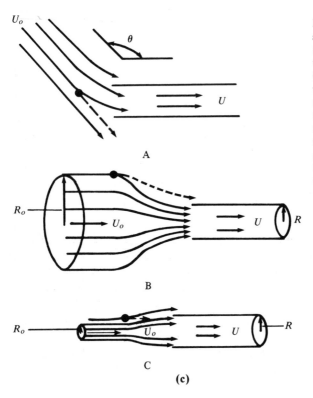

Figure 7.9. Types of anisokinetic sampling. (a) Probe not aligned with flow; $C_{sample} < C_{actual}$. (b) Sample velocity greater than stream velocity; $C_{sample} < C_{actual}$; $C < C_0$. (c) Sample velocity less than stream velocity; $C_{sample} > C_{actual}$; $C > C_0$.

sample concentration which would be less than actual. In case (b), when the collection velocity is greater than stream velocity, some particles, because of their inertia, may fail to follow streamlines and therefore will not be collected, giving a sample concentration that is less than actual. Finally, in case (c), the collection velocity would be less than the stream velocity—the opposite of case (b)—and the sample concentration might then be greater than actual.

If the probe is aligned with the flow, and the sample velocity is equal to the stream velocity, sampling is said to be *isokinetic,* and the sample as collected should match actual concentration. If sampling velocity differs from stream velocity, known as *anisokinetic* sampling, particle inertia can give rise to errors in the measured concentrations.

Error from anisokinetic sampling can be investigated theoretically (Davies, 1966). Suppose an aerosol of concentration c_o flowing with a velocity of u_o is drawn with a velocity u into a sampling tube of radius R (assume infinitely thin tube walls, sample collected parallel to flow). Gas streamlines entering the tube at some distance away are confined within a cylinder of radius R_o such that

$$\pi R^2 u = \pi R_o^2 u_o$$

If $u_o > u$, then $R > R_o$ and the streamlines diverge (Fig. 7.9c). All particles moving within the circular cross-section πR_o^2 enter the tube. The number entering/sec $= \pi R_o^2 u_o c_o$, where c_o is the stream concentration. Streamlines from the annulus which lies between the cylinders of radii R_o and R will pass outside the sampling tube. However, a fraction of particles from this space may enter the tube because of their inertia. If α is this fraction, then the number entering per second and from the annulus is equal to $\alpha\pi(R^2 - R_o^2)c_o u_o$. The sum of the two fluxes, that from the center and that from the annulus, is the total flux entering the tube, equal to $\pi R^2 cu$ where c is the number concentration in the sample:

$$\pi R_o^2 u_o c_o + \alpha\pi(R^2 - R_o^2)\,c_o u_o = \pi R^2 cu \tag{7.32}$$

Hence

$$\frac{c}{c_o} = 1 - \alpha + \frac{\alpha u_o}{u} \tag{7.33}$$

When $u > u_o$, then $c < c_o$. When $u < u_o$, then $c > c_o$.

The factor α varies from zero for small particles to one for large ones. The exact value of α is not known. A rough estimate for α has been proposed by C. N. Davies (1966):

$$\alpha = \frac{2\,\mathrm{Stk}}{1 + 2\,\mathrm{Stk}}$$

where

$$\mathrm{Stk} = \frac{\tau u_o}{R} \tag{7.34}$$

This estimate, although rough, does predict the following experimental observations:

1. $c/c_o = 1$ when $u_o = 0$
2. $c/c_o = 1$ when $u_o > 0$ but $\mathrm{Stk} = \ll 1$
3. $c/c_o = u_o/u$ when $\mathrm{Stk} = \infty$
4. $c/c_o = 1$ when $u = u_o$

The effect of anisokinetic sampling is plotted in Fig. 7.10. This analysis indicates that isokinetic sampling is not necessary when very small particles are sampled. Sometimes, in an effort to be precise, isokinetic sampling will be required when it is known that all particles to be sampled are well below 1 μm in diameter. Examples would be sampling of fumes or aerosols formed from condensation processes. The result is much added complexity and usually cost, without any increase in the accuracy of the result.

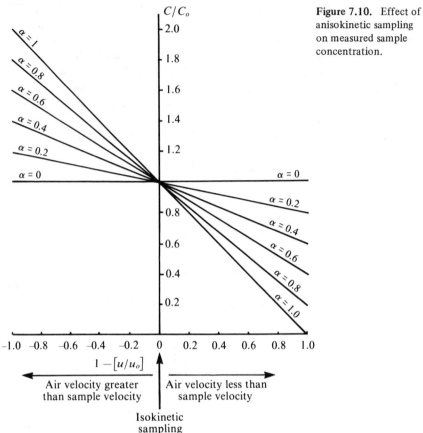

Figure 7.10. Effect of anisokinetic sampling on measured sample concentration.

EXAMPLE 7.11

It is desired to sample fume particles ($d = 0.1$ μm) that are emitted from a stack at a velocity of 100 cm/sec and a temperature of 200°C. (Assume a particle density of 1 g/cm^3 and a sample probe diameter of 1 cm.) Determine the sampling error when the sampling velocity is 0.01 of the stream velocity.

$$\text{Stk} = \frac{\tau v_o}{R}$$

At 200°C

$$\text{Viscosity} = 1.83 \times 10^{-4} \left[\frac{200 + 273}{20 + 273} \right]^{0.5}$$

$$= 2.33 \times 10^{-4} \text{ poises}$$

$$\lambda = 0.065 \times \left[\frac{200 + 273}{273}\right]$$

$$= 0.113 \; \mu m$$

$$\tau = \frac{1}{18} \frac{d^2}{\mu} \rho_p C_c$$

$$\tau = \frac{1}{18} \left(\frac{(0.1 \times 10^{-4})^2}{2.33 \times 10^{-4}}\right)(1)\left[1 + \frac{0.226}{0.1}(1.257)\right]$$

$$\tau = 9.14 \times 10^{-8} \; \text{sec}$$

$$\text{Stk} = \frac{(9.14 \times 10^{-8})(100)}{0.5} = 1.83 \times 10^{-5}$$

$$\alpha = \frac{2\,\text{Stk}}{1 + 2\,\text{Stk}} = 3.65 \times 10^{-5}$$

$$\frac{c}{c_o} = 1 - \alpha + \alpha \frac{u_o}{u} = 1 - (3.65 \times 10^{-5})\left(1 - \frac{100}{1}\right)$$

$$\frac{c}{c_o} = 1.0036$$

that is, there will be a negligible increase in sample concentration.

RESPIRABLE SAMPLING

For the purpose of estimating the toxic dose of an aerosol, the respiratory system can be divided up into a number of functional regions. These are:

1. Alveolar region (for both nose and mouth breathing)
2. Tracheo-bronchial tree (for both nose and mouth breathing)
3. (a) Oral cavity, pharynx and larynx (for mouth breathing)
   (b) Nasopharynx, pharynx and larynx (for nose breathing)
4. Ciliated nasal passages (for nose breathing)
5. Anterior uncilated nares (for nose breathing)

Regional deposition is dependent on the aerodynamic properties of the particles, usually described in terms of the aerodynamic particle diameter, airway dimensions, and such respiratory characteristics as flow rate, breathing frequency, and tidal volume.

A number of models have been developed to attempt to predict respiratory deposition, especially in the lower airways—the alveolar region. Experimental studies have tended to confirm the validity of the models, recognizing that there is much individual variation and thus a great spread in the results. The results have been clear enough, however, to indicate that all else being equal, deposition of particles in the lungs is greatly influenced by particle size and particle density.

In many cases the dose from airborne toxic materials is dependent on regional deposition in the lungs. A good estimate of this dose is possible if the size distribution of the aerosol is known. For this reason it is important to know mass concentrations within various size fractions. This information can be obtained by: (1) carrying out a size distribution analysis of the airborne aerosol; or (2) carrying out a size distribution analysis of the collected sample; or (3) separating the aerosol into size fractions corresponding to anticipated regional deposition during the process of collection.

With mass respirable sampling, an attempt is made to separate the aerosol into two fractions representing the mass that would be deposited in the alveolar region and the mass that would not be deposited in this region. To do this it is necessary to define the size distribution of particles that are deposited in the alveolar region. This material is defined as "respirable dust."

There are several definitions of respirable dust (Lippmann, 1970). In 1952 the British Medical Research Council (BMRC) defined the respirable fraction in terms of the terminal settling velocity (free falling speed), by the equation

$$\frac{c}{c_o} = 1 - \frac{v}{v_c} \tag{7.35}$$

where c is the concentration of particles of falling speed v or less, c_o the total concentration, and v_c is a constant equal to twice the terminal settling velocity in air of a unit density sphere having a diameter of 5 μm. This definition was considered to be unsatisfactory in the United States because it was tied to terminal settling velocities.

In 1961 the U.S. Atomic Energy Commission (AEC) established a standard defining "respirable dust" as that portion of the inhaled dust that penetrates to the nonciliated portions of the lung. This application of the concepts of respirable dust was intended *only* for "insoluble" particles exhibiting prolonged retention in the lung, and not for soluble particles, nor for those which are primarily chemical intoxicants (Aerosol Tech. Committee, 1970).

Respirable dust was defined as follows with sizes in terms of aerodynamic diameters:

Size (μm)	10	5	3.5	2.5	2
Respirable, %	0	25	50	75	100

In 1968 the American Conference of Governmental Industrial Hygienists (ACGIH) defined "respirable dust" as follows:

Aerodynamic diameter (μm)	10	5.0	3.5	2.5	2.0
% passing selector	0	25	50	75	90

This definition of respirable dust is almost identical with that of the AEC, differing only for a 2-μm aerodynamic diameter particle. Lippman (1970) points out that this difference appears to be a recognition by the ACGIH of the characteristics of real particle separators.

Although there are some differences in the three definitions, under field conditions samples collected using instruments designed to meet any of these three criteria should be comparable.

EXAMPLE 7.12

Compare the definitions of respirable dust as given by BMRC, AEC, and ACGIH.

Neglecting C_c, terminal settling velocities for other sizes of unit density spheres can be estimated from

$$v_{g_d} = v_{g_5} \times \left(\frac{d}{5}\right)^2$$

so Eq. (7.35) can be written as

$$\frac{c}{c_o} = 1 - \frac{v_{g_5} \times \left(\frac{d}{5}\right)^2}{2 \times v_{g_5}} = 1 - \frac{d^2}{50}$$

when d is expressed in μm.

% respirable	Aerodynamic diameter, μm				
	2	2.5	3.5	5	10
BMRC	92	88	76	50	0
AEC	100	75	50	25	0
ACGIH	90	75	50	25	0

Figure 7.11 shows a plot of these definitions.

PROBLEMS

1. Using simple impaction theory, estimate the minimum particle diameter that will be collected with 100% efficiency by a plate placed at a right angle to the flow and 4 in. in front of a 4-in. diameter duct out of which air containing an aerosol is flowing at a rate of 100 cfm.

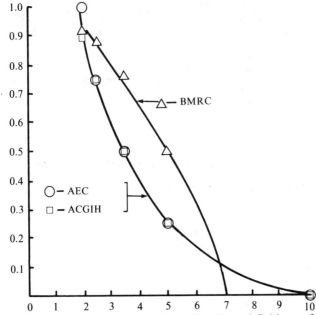

Figure 7.11. Pictorial representation of various definitions of respirable dust.

2. In the first stage of the Lundgren impactor, air issues at a flow of 85 lpm through a number of round jets 0.82 cm in diameter at a Reynolds number of 3700.
   a. Determine the number of jets in the first step of the impactor.
   b. Using the data in Fig. 7.4a, estimate the effective cutoff aerodynamic diameter for this stage.

3. What is the minimum particle diameter collected with 100% efficiency by a cyclone precollector of a mass respirable sampler? Assume $R = 0.5$ cm, $B_c = 0.25$ cm, $Q = 1.7$ lpm and $N_t = 5$, $\rho = 1$ g/cm^3, opening is square.

4. A 1-in. diameter tube is used to collect a stack sample from a stack in which air is flowing at a velocity of 30 fps. The sampling pump available can pump only at a rate of 1 cfm. Estimate the error in sampling for
   a. 10-μm-diameter spheres with $\rho = 2$ g/cm^3
   b. 0.1-μm-diameter spheres with $\rho = 10$ g/cm^3

5. If we assume that the critical Stokes' number for a 1-in.-diameter circular orifice is 1/8 and that the pressure drop for an orifice and collecting plate in series in milliatmospheres is $\Delta P = 1.25 \times 10^{-3}\rho_m U$, where ρ_m = fluid density, g/cc, and U = orifice gas velocity, cm/sec, determine the diameter of the smallest size unit density spherical particle that can be collected at a pressure drop of 2 in. of water.

6. A high-volume sampler has an air flow rate of 30 cfm. Design a horizontal elutriator (settling chamber) that could be placed upstream from the sampler to eliminate those particles greater than 10 μm in diameter. Assume a particle density of 2.3. What size particles would be reduced by a factor of 50% in this unit?

7. The Andersen Sampler is a 6-stage circular jet impactor. Each stage has 400 jets of a certain diameter. For stage 3, the jet diameter is 0.028 in. For stage 4, the jet diameter is 0.021 in. Assuming a flow rate of 1 cfm and particles of unit density, determine the range of particle diameters removed by the stage 4 impaction surface. Use the impaction data given in Fig. 7.4b.

EIGHT

BROWNIAN MOTION AND SIMPLE DIFFUSION

There are two principal ways that extremely small aerosol particles can be removed from an aerosol. The particles can collide with other particles and grow into ones large enough to be removed by gravity or aerodynamic forces (impaction, centrifugal, etc.), or they can migrate to surfaces, stick to those surfaces, and thus be removed.

The process by which these particles migrate, either to a surface or to one another, is called diffusion, and their motion is described as Brownian motion. Diffusion is important in aerosol studies because it represents the major dynamic effect acting on very small particles ($d < 0.1 \ \mu m$) and must be considered when the dynamics of these small particles are studied.

BROWNIAN MOTION

Small particles suspended in a gas undergo random translational motion because they are being buffeted by collisions with swiftly moving gas molecules. This motion appears almost as a vibration of the ensemble of particles, although there is a net displacement with time of any given particle. Observation of this motion in a liquid was first made in 1828 by the British naturalist Robert Brown (1828) and the phenomenon thus has been called *Brownian motion* (also known as Brownian movement). Bodaszewski (1881) studied the Brownian motion of smoke particles and other suspensions in air and likened these movements to the movements of gas molecules as postulated by the kinetic theory. The principles governing Brownian motion are the same whether the particles are suspended in a gas or in a liquid.

FICK'S LAWS OF DIFFUSION

When particles are uniformly dispersed in a gas, Brownian motion will change the position of the individual particles but will not change the overall particle distribution. When the particles are not uniformly dispersed, Brownian motion tends eventually to produce a uniform concentration throughout the gas, the particles moving away from areas of high concentration to regions of low concentration. This process, known as particle diffusion, follows the same two general laws that also apply to molecular diffusion, known as Fick's two laws of diffusion.

Fick's first law of diffusion states that the concentration of particles crossing unit area in unit time, J, is proportional to the concentration gradient normal to the unit area, dc/dx. The constant of proportionality, D, is known as the diffusion coefficient. Symbolically, for the current through a plane set at right angles to the x-direction,

$$J = -D\frac{dc}{dx} \tag{8.1}$$

EXAMPLE 8.1

Particles move by diffusion across a gap 2 cm wide. If the concentration on the left-hand side of the gap is such that it is always 10 times the concentration on the right-hand side (the right-hand side concentration being 10^6 p/cm^3) and 100 particles per second per square centimeter cross the gap, determine the value of D, the diffusion coefficient.

$$J = -D\frac{dc}{dx}$$

$$\frac{dc}{dx} = \frac{1 \times 10^6 - 10 \times 10^6}{2} = \frac{(1-10) \times 10^6}{2} = -4.5 \times 10^6$$

$$D = -\frac{J}{(dc/dx)} = \frac{100}{4.5 \times 10^6} \frac{\dfrac{p}{cm^2 - sec}}{\dfrac{p}{cm^3} \cdot \dfrac{1}{cm}}$$

$$D = 2.22 \times 10^{-5} \ cm^2/sec$$

Notice that the units of D are centimeters squared per second (in cgs units).

Fick's second law represents the time-dependent case in which the change in concentration of an aerosol with respect to time at a point in space is proportional to the divergence of the concentration gradient at that point, the constant of proportionality again being D, the diffusion coefficient (Jost, 1952).

Symbolically this is

$$\frac{\partial c}{\partial t} = D\nabla^2 c \tag{8.2}$$

The term ∇^2 is the Laplacian operator, which in Cartesian coordinates (given by x, y, and z) is

$$\nabla^2 = \frac{\partial^2}{\partial x^2} + \frac{\partial^2}{\partial y^2} + \frac{\partial^2}{\partial z^2} \tag{8.3a}$$

in spherical coordinates (given by r, θ, and ϕ) is

$$\nabla^2 = \frac{\partial^2}{\partial r^2} + \frac{2}{r}\frac{\partial}{\partial r} + \frac{1}{r^2\sin^2\theta}\frac{\partial^2}{\partial\phi} + \frac{1}{r^2}\frac{\partial^2}{\partial\phi^2} + \frac{1}{r^2}\cot\theta\frac{\partial}{\partial\theta} \tag{8.3b}$$

and in cylindrical coordinates (given by r, θ, and z) is

$$\nabla^2 = \frac{\partial^2}{\partial r^2} + \frac{1}{r}\frac{\partial}{\partial r} + \frac{1}{r^2}\frac{\partial^2}{\partial\theta^2} + \frac{\partial^2}{\partial z^2} \tag{8.3c}$$

Thus in Cartesian coordinates Fick's second law of diffusion would be written

$$\frac{\partial C}{\partial t} = D\left(\frac{\partial c^2}{\partial x^2} + \frac{\partial^2 c}{\partial y^2} + \frac{\partial^2 c}{\partial x^2}\right) \tag{8.4}$$

These equations, with appropriate boundary conditions, permit in theory the solution of any aerosol problem involving pure diffusion.

Early investigators using a liquid medium found that a particle in Brownian motion moves with uniform velocity (Svedberg, 1909), that smaller particles move more rapidly than larger ones (Exner, 1867), that particles travel more rapidly as the viscosity of the medium decreases, and that at constant viscosity the amplitude of the motion is directly proportional to the absolute temperature (Seddig, 1908). These observations are consistent with a theory of Brownian motion developed by Albert Einstein (1956) in 1905 and 1906.

EINSTEIN'S THEORY OF BROWNIAN MOTION

Consider a cylinder of unit cross-sectional area in which diffusion of particles is taking place along the axis of the cylinder in a single direction. Within the cylinder are two membranes, E and E', a distance of x from one end and a distance dx apart, as shown in the sketch in Fig. 8.1.

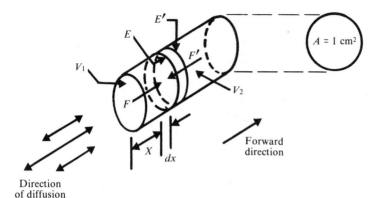

Figure 8.1. Sketch of imaginary cylinder.

Diffusion in the cylinder gives rise to a force from the particles (which could be likened to an osmotic force) acting on the two membranes. The force acting on E is F, whereas the resisting force acting on E' in the opposite direction is F'. The resultant of these forces is $(F - F')$. This force acts on the cylinder volume $A\, dx$, but A, the cross-sectional area of the cylinder, is equal to 1. Thus the force per unit of enclosed volume is $(F - F')/dx$. This is also equal to the osmotic pressure gradient within the enclosed volume, dp/dx, that is, the pressure gradient between the planes E and E'. Letting ΣF_D be the osmotic force per unit volume, we have

$$\Sigma F_D = \frac{(F - F')}{A\, dx} = \frac{-dp}{dx} \tag{8.5}$$

The osmotic pressure of a solute in a solvent is given by the expression

$$p = nRT \tag{8.6}$$

where p is the osmotic pressure, R the gas constant, T the absolute temperature, and n the number of particles expressed as gram molecules per unit volume.

Differentiating with respect to n gives

$$dp = RT\, dn \tag{8.7}$$

which, when substituted in Eq. (8.5) gives, for the osmotic force,

$$\Sigma F_D = -RT \frac{dn}{dx} \tag{8.8}$$

This is the osmotic or diffusional force acting on all the particles per unit volume. If there are n gram-molecules of particles present in the unit volume, then the actual number of particles in the unit volume is determined from the product of n and Avogadro's number, N_A. Since $nN_A = c$ is the number of particles per unit volume, the force acting on each particle, F_D, is

$$F_D = \frac{\Sigma F_D}{nN_A}$$

The resistance offered to the motion of a spherical particle by the medium in which it is moving is given by Stokes' law for small values of the Reynolds number. Equating this resistance with F_D gives

$$F_D = -\frac{RT\frac{dn}{dx}}{nN_A} = 3\,\pi\mu dv/C_c \qquad (8.10)$$

Then, rearranging terms gives

$$N_A nv = \frac{-RT}{N_A}\,\frac{C_c}{3\pi\mu d}\,\frac{d(nN_A)}{dx}$$

or

$$cv = -\frac{RT}{N_A}\,\frac{C_c}{3\pi\mu d}\,\frac{dc}{dx} \qquad (8.11)$$

The product cv represents a diffusion current, that is, the number of particles crossing a unit area in unit time. But Fick's first law of diffusion states that the diffusion current is proportional to the concentration gradient, the constant of proportionality being the diffusion coefficient D. Thus the diffusion coefficient for an aerosol particle is, from Eq. (8.11),

$$D = \frac{RT}{N_A}\,\frac{C_c}{3\pi\mu d} = kT\,\frac{C_c}{3\pi\mu d} \qquad (8.12)$$

where k is the Boltzmann constant.

A new term, particle mobility, B, can be defined. Mobility represents the velocity given to a particle by a constant unit driving force, and is

$$B = \frac{C_c}{3\pi\mu d} \qquad (8.13a)$$

which for a sphere of mass m becomes

$$B = \frac{\tau}{m} \qquad (8.13b)$$

Then the diffusion coefficient is just

$$D = BkT = \frac{\tau}{m}kT \qquad (8.14)$$

Table 8.1. Diffusion coefficients of spheres of various sizes at normal temperature and pressure.

Diameter, cm	D cm^2/sec
10^{-6}	5.24×10^{-4}
10^{-5}	6.82×10^{-6}
10^{-4}	2.74×10^{-7}
10^{-3}	2.38×10^{-8}

EXAMPLE 8.2

Determine the diffusion coefficient of a cigarette smoke particle (spherical shape, $d = 0.25 \ \mu m$, $\rho = 0.9$ gm/cm^3). Assume $T = 20°C$.

$$C_c = 1 + \frac{2\lambda}{d}(1.257) = 1 + \frac{0.14}{0.25}(1.257) = 1.70$$

$$B = \frac{C_c}{3\pi\mu d} = \frac{1.70}{(3)(\pi)(1.83 \times 10^{-4})(0.25 \times 10^{-4})} = 3.95 \times 10^7 \ cm/sec$$

$$D = BkT = (3.95 \times 10^7)(1.38 \times 10^{-16})(293)$$

$$D = 1.60 \times 10^{-6} \ cm^2/sec$$

The diffusion coefficient has units in the cgs system of cm^2/sec and is a function only of particle diameter and gas viscosity for a given temperature (if the Cunningham correction factor can be neglected). At normal conditions of temperature and pressure a 1-μm diameter particle has a diffusion coefficient of 2.74×10^{-7} cm^2/sec, about 10^6 times smaller than the diffusion coefficient for a typical gas molecule. Diffusion coefficients for other particle sizes are given in Table 8.1.

BROWNIAN DISPLACEMENT

When a particle moves in Brownian motion the chance that it will ever return to its initial position is negligibly small. Thus there will be a net displacement with time of any single particle, even though the average displacement for all particles is zero. For example, during a short time interval one particle may move a distance s_1, another a distance s_2, and so on. Some of these displacements will be positive, some negative, some up, some down, but with equilibrium conditions the sum of the displacements will be zero. It is possible to estimate the displacement of any particle in terms of its root mean square displacement.

Suppose for simplicity particles are considered to move forward or backward along only a single axis. Particles moving forward are those considered to have positive velocities. We can denote the root mean square displacement of the forward-moving particles as s, this displacement taking place over the time interval t. During this time interval, on the average, only those particles lying a distance of s or less from a plane E will pass through it. The total number of particles displaced per unit area through E is $(1/2)c_1 s$, where c_1 is the mean concentration of particles within the volume V_1, lying to the left of plane E.

By the same arguments, the concentration of particles going the other way from volume V_2 (on the right of plane E) is $(1/2)c_2 s$. Thus the net flow from left to right is $1/2\,(c_1 - c_2)s$.

However, for very small values of s we can write

$$\frac{dc}{dx} = \frac{c_2 - c_1}{s} \tag{8.15}$$

the definition of a differential. Then

$$c_1 - c_2 = -s\frac{dc}{dx} \tag{8.16}$$

and the net flow can be written as $(1/2)s(dc/dx)$. In a unit of time the quantity J diffusing through a unit area of plane E is

$$J = -\frac{1}{2}\frac{s^2}{t}\frac{dc}{dx} \tag{8.17}$$

But Eq. (8.17) again resembles Fick's first law of diffusion, giving another expression for D, the diffusion coefficient. In this case

$$D = \frac{s^2}{2t} \tag{8.18}$$

The mean square displacement is then

$$s^2 = 2\,Dt \tag{8.19}$$

When movement in three dimensions is considered, the displacement over any given time period will be less than the one-dimensional case, since during part of the elapsed time the particle moves at right angles to the direction of interest. For three-dimensional motion the mean square displacement is

$$s^2 = \frac{4}{\pi}\,Dt \tag{8.20}$$

EXAMPLE 8.3

An aerosol made up of 0.25-μm-diameter smoke particles is collected in a spherical flask 5 cm in diameter. How long will it take, on the average, for a particle to travel from the center of the flask to its outer edge?

From Ex. 8.2, D for a 0.25-μm spherical particle is 1.60×10^{-6} cm^2 sec

$$s = \sqrt{\frac{4}{\pi} Dt}$$

$$2.5 = \sqrt{\frac{4}{\pi} (1.60 \times 10^{-6})t}$$

$$t = 3.07 \times 10^6 \text{ sec}$$

$$= 35.5 \text{ days}$$

This problem illustrates that particles move relatively short distances by diffusion. Thus diffusion is important only when considering particles in very small volumes, close to surfaces, or when particle size is so small that the value of D approaches molecular diffusion coefficients.

BROWNIAN MOTION OF ROTATION

Particles comprising an aerosol move randomly in Brownian motion because of the gas molecules impacting on them. The random nature of the molecules striking the particles can also cause the particles to rotate, this Brownian rotation being described by the equation (Fuchs, 1964)

$$\bar{\theta}^2 = 2 kT B_\theta t. \tag{8.21}$$

The term $\bar{\theta}^2$ is the mean square angle of rotation of the particle about a given axis in the time t. For a spherical particle Fuchs, 1964, gives the rotational mobility, B_θ, as

$$B_\theta = \frac{1}{\pi \mu d^3} \tag{8.22}$$

so that Eq. (8.21) becomes

$$\bar{\theta}^2 = \frac{2kT}{\pi \mu d^3} t \tag{8.23}$$

EXAMPLE 8.4

Determine the average number of revolutions a 5-μm spherical particle will make per minute in air at 20°C.

$$\bar{\theta}^2 = \frac{2kT}{\pi\mu d^3}\, t = \frac{2(1.38 \times 10^{-16})(293)}{(\pi)(1.83 \times 10^{-4})(5 \times 10^{-4})^3}\, (60)$$

$$\bar{\theta}^2 = 67.52$$

$$\sqrt{\bar{\theta}^2} = 8.22 \text{ radians}$$

$$= 1.31 \text{ rpm}$$

For smooth spherical particles Brownian rotation is of no interest because it produces no observable effect. For particles with some irregularities Brownian rotation produces the twinkling effect which is often observed when a beam of light is passed through a cloud of particles. Although Eq. (8.23) was derived for spheres, it can also be applied to isometric particles. For particles smaller than about 1 μm in diameter, the frequency of rotation is faster than the eye can see. On the other hand, with particles having diameters greater than 20μm or so Brownian rotation is very slow. Thus the twinkling normally observed in a cloud arises only from particles whose diameters lie roughly between 1 μm and about 20 μm, but since particles smaller than about 10 μm diameter cannot be seen with the unaided eye, the actual range of sizes of twinkling particles is very small.

"BAROMETRIC" DISTRIBUTION OF PARTICLES

One consequence of kinetic theory is that particles will have the same average translational energy as molecules when the gas is in equilibrium. Thus it is possible to compute the average velocity of a particle as it moves in Brownian motion. Denoting this velocity as v_0,

$$\frac{1}{2}mv_0^2 = \frac{3}{2}kT \tag{8.24}$$

where m is the mass of the particle and $3\,kT/2$ is the average energy of the particle in the gas. Rearranging terms gives

$$v_0 = \sqrt{\frac{3kT}{m}} \tag{8.25}$$

exactly the same as Eq. (3.10), the equation for the root mean square velocity of a gas molecule. Aerosol particles, in their random motion, follow a Maxwell-Boltzmann velocity distribution similar to the molecules. But if they behave similarly to gas molecules, they should also be distributed vertically in equilibrium according to the barometric or Boltzmann equation. This indeed appears to be the case. Monodisperse particles which do not coagulate will be distributed vertically according to the expression

$$c = c_0 \exp\left(\frac{-mgZ}{kT}\right) \qquad (8.26)$$

where Z is the height above some reference point at which the concentration c is measured and c_0 is the concentration of particles at the reference height. This effect is of no importance for particles larger than 0.3 μm. For particles with 0.1-μm diameter, at equilibrium essentially all particles will be contained in a band approximately 0.8 mm thick above a given surface. With particles of 0.01-μm diameter, the band width is approximately 50 cm. Thus many very small particles (less than 0.1-μm diameter) will never be removed by sedimentation on their own accord, being constantly buffeted upward by Brownian motion. Their removal from air must be carried out by some other mechanism.

EXAMPLE 8.5

Extremely fine polonium-210 particles (0.01-μm diameter $\rho = 9.4$ g/cm^3) are spilled on a laboratory bench. Assuming a barometric distribution is established, at what height will $c/c_0 = 0.1$?

$$\frac{c}{c_0} = \exp\left(-\frac{mgZ}{kT}\right)$$

$$\ln\frac{c}{c_0} = \frac{-mgZ}{kT}$$

$$Z = \frac{-kT\ln\dfrac{c}{c_0}}{mg}$$

$$= \frac{-(1.38 \times 10^{-16})(293)(ln(0.1))}{\dfrac{\pi}{6} \times 10^{-18}(9.4)\,(980)}$$

$$Z = 19.30 \text{ cm}$$

Very fine particles will migrate over a surface, possibly as a result of "barometric" resuspension.

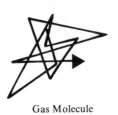

Figure 8.2. Gas molecule trajectory compared to aerosol particle trajectory.

Gas Molecule

Aerosol
particle

EFFECT OF AEROSOL MASS ON THE DIFFUSION COEFFICIENT

Equation (8.12) indicates that the diffusion coefficient of an aerosol particle is independent of particle density, and hence is independent of particle mass. But is this really so? Since particle mass is so much greater than molecular mass, and the particles are continually undergoing bombardment by the molecules, one would expect changes in the direction of the particle to be gradual, compared to the rapid changes in direction with molecular diffusion. But if this is true, then particle momentum (mass) should be considered in the particle diffusion coefficient equation.

Two-dimensional trajectories of a typical gas molecule and a typical aerosol particle can be compared in Fig. 8.2. The molecule shows sharp changes in direction, each change occurring when it strikes another molecule. As discussed in Chapter 3, the average distance between hits is defined as the mean free path of the molecule. For the particle a hit by a single molecule does not appreciably affect its motion; therefore its path is not characterized by sharp changes in direction, but by smooth curves representing the combined effect of hits by many molecules.

The problem can be treated by considering the average particle displacement under the influence of a force whose magnitude and direction vary in a random fashion but whose average magnitude is equal to zero. According to Fuchs (1964), the mean square displacement of an average particle considered in this way is

$$\bar{s}^2 = \frac{2}{3}\, \bar{v}^2\, \tau [t - \tau(1 - e^{-t/\tau})] \tag{8.27}$$

The term $\bar{v}^2$ is equal to $3kT/m$, the mean square Boltzmann velocity. In terms of the diffusion coefficient, Eq. (8.27) becomes

$$\bar{s}^2 = 2D[t - \tau(1 - e^{-t/\tau})] \tag{8.28}$$

When $t \gg \tau$, Eq. (8.28) reduces to Eq. (8.19), an expression for the displacement of a particle at constant velocity. Since our observation times will generally always be greater than τ, we can conclude that in most instances particle inertia can be neglected when considering particle diffusion.

EXAMPLE 8.6

Find the ratio of τ/t such that the root mean square displacement estimated considering particle inertia (Eq. 8.28) is 10% less than the estimate when inertia is not considered (Eq. 8.19).

$$\frac{\text{Eq. (8.28)}}{\text{Eq. (8.19)}} = \frac{2D[t - \tau(1 - e^{-t/\tau})]}{2 Dt} = 0.90$$

$$= 1 - \frac{\tau}{t} (1 - e^{-t/\tau}) = 0.90$$

$$\frac{\tau}{t} (1 - e^{-t/\tau}) = 0.1$$

By trial and error,

$$\frac{\tau}{t} = 0.1$$

that is, when $t \geq 10\tau$, this correction is unnecessary.

AEROSOL APPARENT MEAN FREE PATH

Since aerosol particles are continually undergoing molecular bombardment, their paths are smooth curves rather than segments of straight lines. It still is possible to define an *apparent mean free path* for the aerosol particles (Fuchs, 1964). This is the distance traveled by an average particle before it changes its direction of motion by 90°. Apparent mean free path represents the distance traveled by an average particle in a given direction before particle velocity in that direction equals zero. But this is just the stop distance.

At any time, a particle may be considered to be moving in a specific direction with a velocity $v = \sqrt{8 kT/\pi m}$. From a definition of the stop distance, the pseudo mean free path l_B, is

$$l_B = \tau v = \tau \sqrt{8 kT/\pi m} \tag{8.29}$$

At normal pressure and temperature, l_B reaches a minimum at an aerosol particle diameter of 5×10^{-6} cm, but increases only by about a factor of 5 for particles two orders of magnitude larger or smaller than this size. Thus the pseudo mean free path is essentially constant over the size range of interest, having a value of about 10^{-6} cm.

EXAMPLE 8.7

Compute the apparent mean free paths for unit density spheres of 0.01-μm, 0.1-μm, and 1-μm diameters. Assume $T = 20°C$.

$$l_B = \pi \sqrt{\frac{8\,kT}{\pi m}}$$

d(μm)	C_c	τ	m
0.01	23.8	7.22 $\times$ 10^{-9}	5.24 $\times$ 10^{-19}
0.1	3.02	9.15 $\times$ 10^{-8}	5.24 $\times$ 10^{-16}
1.0	1.18	3.57 $\times$ 10^{-6}	5.24 $\times$ 10^{-13}

$$l_B = \tau \sqrt{\frac{8(1.38 \times 10^{-16})(293)}{\pi m}}$$

$$l_B\,(0.01\;\mu m) = 3.20 \times 10^{-6}\;cm$$

$$l_B\,(0.1\;\mu m) = 1.28 \times 10^{-6}\;cm$$

$$l_B\,(1.0\;\mu m) = 1.58 \times 10^{-6}\;cm$$

PROBLEMS

1. Compute the diffusion coefficient in air of a 5-μm unit density sphere at 20°C.

2. Repeat problem 1 for a 0.5-μm unit density sphere.

3. Repeat problem 2 for temperatures of 0°C and 100°C.

4. Estimate the root mean square displacement for a 2-μm silica dust particle ($\rho = 2.65$ g/cm^3) over a 10-minute period.

5. How long on the average will it take a 0.25-μm cigarette smoke particle (assume sphere, $\rho = 1.0$ g/cm^3) to diffuse
   a. 1 ft
   b. 10 ft.

6. Assuming that a person can distinguish individual flashes of light appearing at a frequency of 5/sec or less, estimate the minimum size of a particle that will appear to twinkle in a beam of sunlight.

PARTICLE DIFFUSION

In particle diffusion the migration or movement of aerosol particles down a concentration gradient is considered. Since particles will always tend to move from regions of high concentration to regions of low concentration (Chapter 3) there will always be a tendency for aerosols to migrate to walls or other surfaces where the concentration, because of deposition, is essentially zero. As discussed in Chapter 8, the range over which this migration occurs is quite small. In those cases where the aerosol is in a fairly confined space to begin with, such as in the lung or in a small sampling tube, loss of the aerosol by diffusion can be significant. In this chapter methods for estimating this loss will be described.

STEADY-STATE DIFFUSION

Consider the case of the diffusion of particles in a gas where the concentration of particles, although varying at different points within a gas, does not change with time. An example is the diffusion of particles from a zone of constant concentration, $c = c_0$, to a wall, where the airborne concentration is assumed to be zero. Suppose it is desired to know the deposition rate of particles on the wall due to diffusion. From Fick's first law, the diffusion current J is

$$J = -D\frac{dc}{dx} \tag{9.1}$$

121

If δ is the distance from the zone of constant concentration to the wall, then the concentration gradient, dc/dx, will be

$$\frac{dc}{dx} = \frac{0 - c_0}{\delta} = -\frac{c_0}{\delta} \qquad (9.2)$$

so that number of particles striking a unit area of the wall in unit time is

$$J = \frac{Dc_0}{\delta} \qquad (9.3)$$

EXAMPLE 9.1

An aerosol flowing through a tube is kept at a constant concentration inside the tube to within 1 mm of the tube wall. If the aerosol is made up of 0.5-μm-diameter spheres and the concentration in the tube is 10^3 particles per cubic centimeter, estimate the wall deposition rate, in particles per square centimeter per second. Assume $T = 20°C$.

$$C_c = 1 + A\frac{\lambda}{d}$$

$$C_c = 1.35$$

$$D = BkT = \frac{C_c}{3\pi\mu d}(kT) = \frac{1.35 \times 1.38 \times 10^{-16} \times 293}{3(\pi)(1.83 \times 10^{-4})(5 \times 10^{-5})}$$

$$D = 6.33 \times 10^{-7} \, cm^2/sec$$

$$J = \frac{D \cdot c_0}{\delta} = \frac{6.33 \times 10^{-7} \cdot 10^3}{0.1}$$

$$J = 6.33 \times 10^{-3} p/cm^2 - sec$$

This is a negligible deposition rate. Unfortunately, it is quite difficult to estimate δ, the concentration boundary layer thickness.

NON-STEADY-STATE DIFFUSION

In Ex. 9.1 the case where the aerosol concentration does not change with time was considered. In many practical situations, however, the aerosol concentration does change with time, possibly as a result of diffusion and subsequent loss of the particles to a wall or other surface. In this event Fick's second law, Eq. (8.2), must be used. Solution of this equation is possible in many cases, depending on the initial and boundary conditions chosen, although the solutions generally take on very complex forms and the actual mechanics involved to find these solutions can be quite tedious. Fortunately, there are several excellent books available which contain large numbers of solutions to the transient diffusion equation (Barrer, 1941; Jost, 1952). Thus in most cases it is possible to fit initial and boundary conditions of an aerosol problem to one of the published solutions. Several commonly occurring examples follow.

Infinite Volume, Plane Vertical Wall

Consider the case of a plane vertical wall that is in contact with an infinitely large volume of aerosol having the same initial concentration throughout. It is desired to estimate the rate of deposition of aerosol on a unit area of wall, assuming that all particles hitting the wall stick to it. The conditions of the problem make it one-dimensional. Letting x be the distance from the wall, Eq. (8.2) becomes

$$\frac{\partial c}{\partial t} = D \frac{\partial^2 c}{\partial x^2} \tag{9.4}$$

since the concentration gradients in the y and z directions are zero. With the initial concentration $c(x, 0) = c_0$ and the boundary condition $c(0, t) = 0$ for $t > 0$, the solution to Eq. (9.4) is

$$c(x, t) = \frac{2c_0}{\sqrt{4\pi Dt}} \int_0^x e^{-\S^2/(4Dt)} d\S \tag{9.5}$$

or

$$c(x, t) = c_0 \operatorname{erf}\left(\frac{x}{\sqrt{4Dt}}\right) \tag{9.6}$$

where erf represents the probability or error function, a tabulated function (see Appendix C). When the argument of this function is small, the function is small and $\operatorname{erf}(0) = 0$. For arguments greater than about 2.6, $\operatorname{erf}(x) = 1$. Steep concentration gradients initially occur close to $x = 0$, gradually decreasing with time (Fig. 9.1).

Recalling that the diffusion current, J, which is the number of particles crossing unit area in unit time, is equal to the diffusion coefficient times the gradient (Fick's first law), evaluating the gradient at $x = 0$, gives the number of particles deposited in the time interval dt

$$J dt = -D \frac{\partial c}{\partial x} dt$$

$$x = 0$$

From Eq. (9.6) and Appendix C,

$$\frac{\partial c}{\partial x} = \frac{2c_0}{\sqrt{\pi}} \cdot \frac{1}{\sqrt{4Dt}} \cdot \exp\left(-\frac{x^2}{4Dt}\right) \tag{9.7}$$

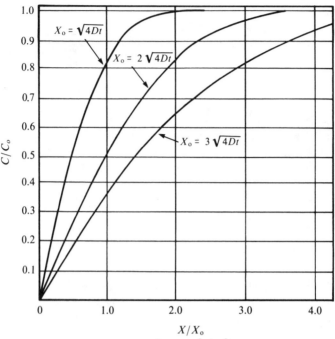

Figure 9.1. Plot of solution of $\partial c/\partial t = D(\partial^2 c/\partial x^2)$ (semi-infinite case).
Solution: $c/c_0 = \text{erf}(x/\sqrt{4Dt})$.

which is to be evaluated at $x = 0$. Then

$$Jdt = dN = c_0 \sqrt{\frac{D}{\pi t}}\, dt \qquad (9.8)$$

at $x = 0$ which on integrating from $t = 0$ to $t = t$ gives N, the number of particles deposited per unit area in the time interval t, i.e.,

$$N = c_0 \sqrt{\frac{4Dt}{\pi}} \qquad (9.9)$$

EXAMPLE 9.2

Estimate the number of 0.1-μm-diameter particles deposited per square centimeter per hour on a wall placed next to a semi-infinite aerosol containing 100 particles/cm^3.

From Table 8.1, D for 0.1-μm particles = 6.82×10^{-6} cm^2/sec

$$N = c_0 \sqrt{\frac{4Dt}{\pi}} = 100\left[\frac{(4)(6.82 \times 10^{-6})(60 \times 60)}{\pi}\right]^{1/2}$$

$$N = 17.68 \text{ particles/cm}^2 - \text{hr}$$

It is interesting to compare Eq. (9.3) and (9.9). Equation (9.3) represents steady-state conditions, in which the concentration a distance away from the wall is always constant, while Eq. (9.9) relates to the case where the concentration near the wall decreases as particles are lost to the wall.

Two Vertical Walls a Distance H Apart

Suppose instead of being semi-infinite, the cloud is contained between two vertical walls spaced a distance H apart. With the same initial conditions as before, and the boundary condition that $c = 0$ at both $x = 0$ and $x = H$ when $t = 0$, then the solution to Eq. (9.4) becomes

$$\frac{c}{c_0} = \text{erf}\left(\frac{x}{\sqrt{4Dt}}\right) - \left[\text{erf}\left(\frac{H+x}{\sqrt{4Dt}}\right) - \text{erf}\left(\frac{H-x}{\sqrt{4Dt}}\right)\right] \qquad (9.10)$$

Figure 9.2 shows a plot of c/c_0 as a function of $(4Dt/H^2)$ for the cases where $x = (H/2)$ and $x = (H/20)$. When H is large compared to x the solution is equivalent to the single wall, infinite medium case.

Figure 9.3 is a plot of c/c_0 versus time, measured at $x = H/2$ for monodisperse particles having a diameter of 1 μm when $H = 2$ cm. Note that there is essentially no change in concentration until sometime after 10^5 sec, and then a

Figure 9.2. Finite case. Change in concentration by diffusion occurring between two walls spaced a distance of H apart; Equation (9.10).

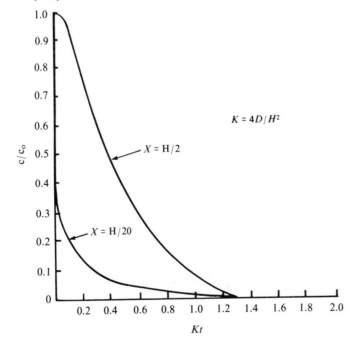

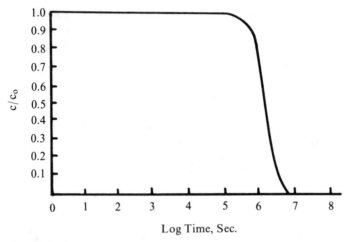

Figure 9.3. Diffusion of monodisperse 1-μm diameter spheres. Concentration measured at center of two plates spaced 2 cm apart.

fairly rapid decrease in concentration takes place. It can be concluded that except for very small particles or very small tubes pure diffusion will have a small to negligible influence on the concentration changes in an aerosol flowing through a tube.

But concentration change, and hence "diffusive" deposition, is observed. Thus there must be an additional mechanism operating which tends to enhance deposition of small particles by "diffusion."

TUBE DEPOSITION

It is possible to make a rough estimate of the diffusional deposition in a tube of radius R by assuming a residence time of $t = L/u$, where u is the velocity in a tube of length L

$$c_{out} = c_{in} - N \cdot \left[\frac{\text{area of tube}}{\text{volume of tube}} \right]$$

$$c_{out} = c_{in} - 2c_{in} \sqrt{\frac{DL}{\pi \mu}} \left[\frac{2RL\pi}{\pi R^2 L} \right]$$

$$\frac{c_{out}}{c_{in}} = 1 - \frac{4}{\sqrt{\pi}} \sqrt{\frac{DL}{uR^2}}$$

(9.11a)

This result is interesting because it indicates that for diffusional deposition with a fixed flow rate, deposition is the same whether one uses a large tube with a low velocity or a small tube with a high velocity. This result is borne out in a more accurate diffusional deposition form given by Gormley and Kennedy (1949), who found that the expression

$$\frac{c_{out}}{c_{in}} = 1 - 2.56 \, \phi^{2/3} + 1.2 \, \phi + 0.177 \, \phi^{4/3} \qquad (9.11b)$$

gave quite acceptable results when compared to experimental data. In Eq. (9.11b) the factor ϕ is given by

$$\phi = \frac{DL}{uR^2}$$

EXAMPLE 9.3

An aerosol made up of 0.1-μm diameter particles ($D = 6.82 \times 10^{-6}$ cm^2/sec) flows through a 1-cm-diameter tube at 15 liters per minute. If the tube is 100 ft long, estimate c_{out}/c_{in}:

a. Using Eq. (9.11a)

$$u = \frac{15,000}{60} \bigg/ \frac{\pi}{4}(1)^2 = 318.3 \text{ cm/sec}$$

$$\frac{c_{out}}{c_{in}} = 1 - \frac{4}{\sqrt{\pi}}\sqrt{\frac{6.82 \times 10^{-6} \times 100 \times 30.5}{318.3 \times (0.5)^2}}$$

$$= 0.964$$

b. Using Eq. (9.11b)

$$\phi = 2.61 \times 10^{-4}$$

$$\frac{c_{out}}{c_{in}} = 0.990$$

A solution for rectangular tubes of varying aspect ratios was recently given by Koch (1982). He showed that for square tubes, deposition approximated the pattern given by Gormley-Kennedy, but was slightly greater.

DIFFUSION IN FLOWING AIR STREAMS – CONVECTIVE DIFFUSION

Thus far only models for diffusion of particles in still or stagnant air have been investigated. Very often, however, the air in which the particles are suspended is not stagnant but has some overall motion. As an example, consider the smoker in a room full of people. Although there may be no perceptible air movement, when a cigarette is lit the odor of tobacco smoke is quickly detectable through-out the room. Even if molecular diffusion coefficients were used to describe the motion of the tobacco smoke particles, transport rates by diffusion are too small to explain the appearance of smoke so quickly in all parts of the room. What occurs is that particles are entrained and transported by the moving air within the room. Convective diffusion describes this phenomenon.

GENERAL EQUATIONS OF CONVECTIVE DIFFUSION

First consider the flux of particles in a fluid through unit area in unit time. Parti-cles can be transported by molecular diffusion or by the moving fluid. For molecular diffusion, from Fick's first law, $J_D = -D$ grad c, whereas if particles are entrained in a moving liquid,

$$J_{CONV} = c\vec{u} \tag{9.12}$$

The total mass flux is the sum of the two fluxes and is expressed as the vector

$$J = c\vec{u} - D \vec{\text{grad}}\, c \tag{9.13}$$

From Eq. (9.13) it can be shown (Levich, 1962) that the time rate of change of concentration — similar to Fick's second law — is

$$\frac{\partial c}{\partial t} = D\nabla^2 c - u\nabla c \tag{9.14}$$

This is the general convective diffusion equation for particles in an isothermal gas when the particles are not subjected to any forces other than the convective motion of the gas and the molecular motion of the gas molecules.

If a volume source is present, such as a gas-phase reaction which produces Q_0 particles per cubic centimeter throughout the volume, Eq. (9.14) becomes

$$\frac{\partial c}{\partial t} = D\nabla^2 C - u\nabla c + Q_0 \tag{9.15}$$

In rectangular Cartesian coordinates, Eq. (9.14) can be written

$$\frac{\partial c}{\partial t} + u_x \frac{\partial c}{\partial x} + u_y \frac{\partial c}{\partial y} + u_z \frac{\partial c}{\partial z} = D\left(\frac{\partial^2 c}{\partial x^2} + \frac{\partial^2 c}{\partial y^2} + \frac{\partial^2 c}{\partial z^2}\right) \tag{9.16}$$

When $u_x = u_y = u_z = 0$, indicating no convective motion of the gas, Eq. (9.16) reverts back to the "pure" diffusion case. The terms, u_x, u_y, and u_z are not necessarily equal, nor are they usually constant, since convective velocities de-crease as a surface is approached. Equation (9.16) thus represents a second-order partial differential equation with variable coefficients. These types of equations are usually quite difficult to solve. However, often it is sufficient to only consider the steady-state solution; that is, the case where $\partial c/\partial t = 0$, indicating that the concentration at any point within the system is not changing with time. Then Eq. (9.16) becomes

$$u_x \frac{\partial c}{\partial x} + u_y \frac{\partial c}{\partial y} + u_z \frac{\partial c}{\partial z} = D\left(\frac{\partial^2 c}{\partial x^2} + \frac{\partial^2 c}{\partial y^2} + \frac{\partial^2 c}{\partial z^2}\right) \qquad (9.17)$$

Convective Diffusion Defined by the Peclet Number

If u_0 is the average velocity in a system where both molecular diffusion and convective diffusion are taking place, L is a characteristic length, and c_0 a representative concentration, then Eq. (9.17) can be put in dimensionless form by making the following substitutions: $U_x = u_x/u_0$, $U_y = u_y/u_0$, $U_z = u_z/u_0$, $C = c/c_0$, $X = x/L$, etc. Equation (9.17) becomes

$$U_x \frac{\partial C}{\partial X} + U_y \frac{\partial C}{\partial Y} + U_z \frac{\partial C}{\partial Z} = \frac{1}{Pe}\left(\frac{\partial^2 C}{\partial X^2} + \frac{\partial^2 C}{\partial Y^2} + \frac{\partial^2 C}{\partial Z^2}\right) \qquad (9.18)$$

where Pe defines the dimensionless ratio

$$Pe = u_0 L/D \qquad (9.19)$$

known as the Peclet number. It is clear that one side of Eq. (9.18) represents molecular motion while the other side represents convective motion. Since all the dimensionless terms in Eq. (9.18) are essentially one (Levich, 1962), the Peclet number describes the relationship between diffusion and convection in a manner similar to the role played by the Reynolds number in fluid flow. When the Peclet number is small, molecular diffusion predominates. When it is large, convective transport predominates and diffusion can be neglected.

EXAMPLE 9.4

Determine the Peclet number for 0.25-μm-diameter spheres being mixed in a room 10 ft wide, 20 ft long and 10 ft in height if air is circulating in the room at a rate of 6 air changes per hour.

$$\text{Volumetric flow rate} = \text{volume} \times \frac{\text{changes}}{\text{hour}} = (10 \times 20 \times 10) \times 6$$

$$= 12{,}000 \text{ cf/hour} = 200 \text{ cfm}$$

Using room cross-sectional area,

$$u_0 = \frac{200}{10 \times 10} = 2 \text{ fpm} = 1.02 \text{ cm/sec}$$

$$D \text{ for } 0.25\text{-}\mu\text{m spheres} = BkT = \frac{C_c}{3\pi\mu d}kT$$

$$D = \frac{\left[1 + 1.257 \dfrac{0.14}{0.25}\right][1.38 \times 10^{-16}][293]}{3\pi(1.83 \times 10^{-4})(0.25 \times 10^{-4})}$$

$$= 1.60 \times 10^{-6} \text{ cm}^2/\text{sec}$$

$$Pe = \frac{u_0 L}{D} = \frac{1.02 \times (20 \times 30.5)}{1.60 \times 10^{-6}}$$

$$Pe = 3.88 \times 10^8$$

Pe is very large; hence convection predominates and diffusion can be ignored.

In this solution 20 ft was chosen for L because we were interested in mixing along the entire length of the room. However, if either the width or height were chosen for L, the resulting conclusion would be exactly the same!

As shown earlier, particle diffusion coefficients are fairly small (in the order of 10^{-4} to 10^{-8} cm^2/sec), resulting in large Peclet numbers unless u_0 (the average velocity in the system) is quite small, or the characteristic length is quite small. In most cases of interest average convective velocities are 0.01 cm/sec or greater, not sufficiently small enough by themselves to ensure a small Peclet number (and hence a diffusion controlled problem). Thus, whether diffusion or convection predominates generally depends solely on the definition of the characteristic length, L. This has already been defined as the length over which the major change in concentration takes place.

Consider air flowing over a flat surface. If the average concentration of particles in the air is c_0 and the concentration at the surface is zero, it is expected that the concentration change from c_0 to 0 would occur mainly near the surface. This has already been shown to be the case for molecular diffusion alone. The distance over which this concentration change occurs is the characteristic length. In other words, molecular diffusion is important in convective diffusion only in the small region close to surfaces. Here it is extremely important, since not only are concentration gradients decreasing sharply, but velocity gradients are also decreasing.

The zones where these gradients occur are often called boundary layers. For example, the aerodynamic boundary layer is the region near a surface where viscous forces predominate. It is of great value in the study of fluid mechanics. Boundary layers exist both with laminar and turbulent flow and may be either laminar or turbulent themselves (Landau and Lifshitz, 1959).

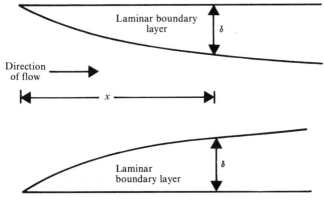

Figure 9.4. Development of laminar boundary layer (not to scale).

Laminar Boundary Layer

With air flowing over a surface, the boundary layer thickness, δ, increases along the surface in the direction of flow (Davies, 1966):

$$\delta = 5 \sqrt{\frac{x \upsilon}{u_0}} \qquad (9.20a)$$

The factor δ represents distance from the surface to the point where 99% of u_0, the mainstream velocity, is reached (see Fig. 9.4). In Eq. (9.20a), x is the distance measured in the direction of flow from the starting point to the point of interest, the kinematic viscosity. If we define a linear Reynolds number, Re_x, as $u_0 x/\upsilon$, Eq. (9.20a) can be written as

$$\delta = \frac{5x}{\sqrt{Re_x}} \qquad (9.20b)$$

For laminar air flow in a tube, when δ approaches the tube radius, Poiseuille flow or a parabolic flow profile is fully developed. This is accomplished by the acceleration of the central portion of the flow. On the other hand, when Re_x exceeds a value lying somewhere between 10^4 and 10^6, the laminar boundary layer becomes so thick that it is no longer stable, and a turbulent boundary layer develops.

EXAMPLE 9.5

Twenty liters of air flow per minute into a 2-in. diameter tube of circular cross section.

    a. Find the boundary layer thickness a distance of 1 cm into the tube.

To determine the boundary layer thickness at 1 cm,

$$\text{Re}_x = \frac{u_0 x}{v} = \frac{4\left(\dfrac{20000}{60}\right)}{\pi(2.54 \times 2)^2} \times \frac{(1)}{0.152} = 16.45/0.152$$

$$= 108.2$$

Knowing Re_x, the thickness δ can be determined.

$$\delta = \frac{5(x)}{\sqrt{\text{Re}_x}} = \frac{5(1)}{\sqrt{108.2}} = 0.48 \text{ cm}$$

b. At what distance into the tube will a parabolic laminar flow profile be fully established?

Let δ = tube radius = 1 × 2.54 cm

$$\delta = 5 \sqrt{\frac{xv}{u_0}}$$

$$\delta^2 = 5^2 \frac{xv}{u_0}$$

$$x = \frac{\delta^2 u_0}{5^2 v} = \frac{(1 \times 2.54)^2(16.45)}{(25)(0.152)}$$

$$x = 27.93 \text{ cm}$$

With fully developed laminar flow, the velocity at any point r in a circular tube of radius R can be expressed by the equation

$$u(r) = u_m \left(1 - \frac{r^2}{R^2}\right) \tag{9.21}$$

where u_m is the maximum center line flow velocity, or twice the average tube velocity.

Turbulent Boundary Layer

A turbulent boundary layer is actually made up of three zones, a viscous or laminar sub-layer immediately adjoining the wall, a buffer zone, and finally a turbulent zone making up the main boundary layer (Schlicting, 1968). Generally speaking, turbulent boundary layers are thicker than laminar boundary layers.

Concentration Boundary Layer

Since both laminar and turbulent boundary layers contain laminar or viscous layers, it would seem logical that diffusion would primarily take place across these regions. If the boundary layer thickness were known, assuming a linear decrease in concentration, Eq. 9.3 could be used to estimate diffusion current. Unfortunately, the point of uniform velocity is not necessarily point of uniform concentration. This is because particles, with their large inertia compared to air, can be carried into laminar boundary regions by mixing as well as by diffusion. The value for δ in Eq. (9.3) will always be less than the equivalent value for the aerodynamic boundary layer thickness, in some cases being only one-tenth or even smaller (Levich, 1962).

Thus there are actually two boundary layers of interest, the aerodynamic boundary layer which is a result of the velocity gradient established at the boundary, and the diffusion or concentration boundary layer resulting from the concentration gradient which exists near the surface.

For turbulence it is convenient to describe particle flux in terms of an eddy diffusion coefficient, similar to a molecular diffusion coefficient. Unlike a molecular diffusion coefficient, however, the eddy diffusion coefficient is not constant for a given temperature and particle mobility but decreases as the eddy approaches a surface. As particles are moved closer and closer to a surface by turbulence, the magnitude of their fluctuations to and from that surface diminish, finally reaching a point where molecular diffusion predominates. As a result, in turbulent deposition, turbulence establishes a uniform aerosol concentration that extends to somewhere within the viscous sub-layer. Then molecular diffusion or particle inertia transports the particles the rest of the way to the surface.

As particle size increases, particles tend to lag behind the eddy motion of the turbulent air. Particle size may be so large that particles are influenced only slightly or not at all by the turbulence. In this case particles will not be deposited by turbulent motion. Smaller particles that follow turbulence, even though they might lag behind, can be deposited by being projected across the boundary layer if the boundary layer thickness is less than the particle stop distance (Sehmel, 1968). Since increasing turbulence tends to increase particle motion, increases in turbulence will tend to enhance particle deposition for a given size particle. On the other hand, at a given level of turbulence (Reynolds number), calculations made by Davies (1965) indicate that there exists a particle size having a maximum rate of deposition, as shown in Fig. 9.5. These deposition rates can be expressed in terms of a "deposition" or "diffusion" velocity.

The Diffusion Velocity

If a concentration boundary layer, δ, can be defined, then the number of particles deposited per unit area in unit time becomes

$$J = \frac{c_0 D}{\delta} \tag{9.22}$$

By dividing J by c_0, a term having the units of velocity results. This function,

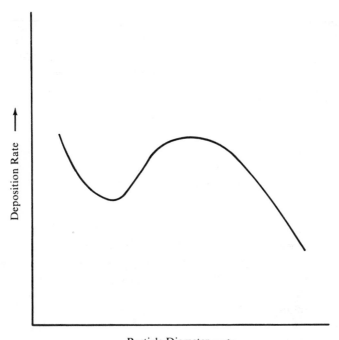

Figure 9.5. Schematic diagram of deposition velocity as a function of particle diameter (Reynold's number fixed).

v_D, is called the diffusion velocity, defined as

$$v_D = \frac{J}{c_0} = \frac{D}{\delta} \tag{9.23}$$

Following this same logic, a diffusion "force" can be defined as

$$F_{\text{diff}} = \frac{3\pi\mu v_D d}{C_c} = \frac{kT}{\delta} \tag{9.24}$$

which is dependent on particle and medium properties only as δ is dependent on these properties.

Application of Diffusion Velocity

Consider a well-mixed ensemble of particles flowing through a tube of radius R with no other factors but diffusion tending to remove the particles from the flow. With diffusion velocity considered as a net movement of particles to the tube surface, in an interval of time of one second there will be $J(2\pi R)$ particles deposited per unit length of tube. In a time, $dt = dx/u_0$ a 1-cm length of aerosol traverses a distance dx. Thus in this time $J(2\pi R)dx/u_0$ particles are removed, and the change in concentration is the number of particles removed divided by the volume from which they are removed, or

$$dc = J(2\pi R)\,dx/u_0 \cdot (1/\pi R^2) = cv_D \times 2dx/u_0 R \qquad (9.25)$$

Integrating, with the initial condition that $c = c_0$ at the entrance of the tube, gives

$$\ln\frac{c}{c_0} = -\left(\frac{2v_D L}{u_0 R}\right) \qquad (9.26)$$

where L is the overall length of the tube. If deposition velocity is a constant for the length of tube considered, it is possible to estimate deposition within the tube from Eq. (9.26).

Conversely, this equation can be used to determine deposition velocities from experimental data.

EXAMPLE 9.6

2-μm unit density spheres are deposited while flowing at a rate of 24 lpm through a 0.21-in.-diameter tube. If the concentration downstream of a 100-cm tube length is 87% of the initial or upstream concentration, estimate the particle deposition velocity.

$$\ln(c/c_0) = -\left[\frac{2v_D L}{u_0 R}\right]$$

$$u_0 = \frac{24 \times 10^3}{(60)\pi\left(\dfrac{0.21 \times 2.54}{2}\right)^2} = 1.79 \times 10^3 \text{ cm/sec}$$

$$R = \frac{0.21 \times 2.54}{2} = 0.267 \text{ cm}$$

$$-0.14 = -0.419\, v_D \qquad v_D = 0.334 \text{ cm/sec}$$

Experimental determinations of v_D are complicated by entrance effects as well as by the effect of gravity, which is usually ignored. As a result only order-of-magnitude accuracy has been achieved from predictions made using the equations of this section. Even so, it should be clear that deposition is largely determined by the properties of the fluid flowing near the wall. Factors such as surface shape or roughness, since they affect this fluid flow, will have a marked effect on deposition, even at low stream velocities.

PROBLEMS

1. Using the barometric equation, compute the height at which 50% of 0.05-μm unit density spheres would be suspended by molecular impacts.

2. Estimate the apparent mean free path of 0.1-μm unit density spheres in air at 20°C, 760 mm pressure.

3. For a 0.01-μm unit density sphere, how short must a diffusion experiment be to have as much as 1% error in distance measurements? How much shorter for a 10% error?

4. A cloud of 0.05-μm spheres is held in a large container. The initial concentration is 10^4 particles/cm^3. After 20 min, what is the aerosol concentration (in particles/cc) 0.1 mm from the wall? How long will it take the aerosol to decrease to a concentration of 10^3 particles/cm^3 1 cm from the wall?

5. A sheet of glass 4 cm by 4 cm square is inserted into a cloud containing 10^5 0.02-μm spherical particles/cm^3. If a microscope is used with a viewing area of 50 μm $\times$ 50 μm to view these particles, and 100 particles are observed per field, what is the average areal density of particles on the glass, and how long must a sample be collected to achieve this density?

6. Mercer (1973) gives an equation for the penetration P of particles flowing in laminar flow through a tube of radius R and length L with a volumetric flow rate (cm^3/sec) as

   $$P = 0.819e^{-1.828\mu} + 0.0975e^{-11.15\mu} + 0.032e^{-28.48\mu} + 0.0157e^{-53.8\mu} + \cdots$$

   where $\mu = 2\pi DL/Q$.
   Compare this equation to that of Gormley and Kennedy (1949) for

   $$\mu = 0.001, 0.005, 0.01, 0.05, 0.1, 0.5$$

7. For a rectangular channel of width B and height H the corresponding equation given by Mercer is

   $$P = 0.9104e^{-1.885\mu} + 0.0531e^{-21.43\mu} + 0.0153e^{-63.23\mu} + 0.0068e^{-124.5\mu} + \cdots$$

   where $\mu = 2DBL/QH$.
   Determine the value of μ for which $P = 0.5$.

8. In the portable diffusion battery of Sinclair (1972) air is flowed through a diffusion battery 1.38 in. in length, 1 3/4 in. in diameter containing 14,500 holes each 0.009 in. in diameter. Samples can be drawn out at different points along the length. Show that the equivalent length (i.e., the length of a battery consisting of one tube) is equal to the actual length times the number of holes.

9. For the diffusion battery of Problem 8, determine the maximum particle diameter which can be collected with 50% efficiency with a total flow through the unit of 1/min. For penetration use the equation

   $$P = 0.819e^{-365\mu} + 0.097e^{-22.3\mu} + 0.032e^{-57\mu} + 0.027e^{-123\mu} + 0.025e^{-750\mu}$$

   where $\mu = \pi Dl/q$,

   l = length of battery, and

   q = volume flow rate per tube cc/sec

AEROSOLS CHARGING MECHANISMS

INTRODUCTION

Up to this point aerosol particles have been considered to be uncharged; that is, electrical forces acting on or between particles were neglected. Most aerosols carry some electrical charge which may be continually transferred between particles, or gained or lost, depending on a number of external factors. The role of electricity in aerosol behavior is not completely understood, even though there is great interest in this particular phenomenon for such diverse reasons as the prevention of dust explosions or better prediction of particle behavior. It was measurement of charge on aerosol particles that gave the first accurate measure of the unit charge of an electron. Electrical forces offer a highly efficient air cleaning method, and the study of very small particles is most conveniently carried out by analyzing their mobility or movement in an electrical field. The possibility of electrostatic propulsion for space vehicles has also generated interest in electrical phenomena of aerosols.

Several electrical properties may be of interest in aerosol studies. These could include the distribution of charges carried by aerosol particles and the velocity of a charged particle in an electric field. This latter property, for example, is important in determining such things as deposition rates or charge transfer rates.

DEFINITION OF FORCE

Suppose a charged, dilute, monodisperse aerosol made up of spherical particles is placed in a uniform electric field, and the movements of the particles making up the aerosol are observed. Some particles will rise, some will fall, and some will remain suspended. From this observation the conclusion can be drawn that the electric field acts as a "field of force" that is superimposed on other forces already present, in this case gavity. However, as seen from the different motions of the particles, their trajectories (whether up or down) are determined by an additional factor as well, in this case the charge carried by each individual particle. Since the magnitude and direction of the electrical force acting on each particle appears to depend not only on the direction and strength of the field but also on the charged state of the particle (including the sign of the charge), a force vector, F_E is defined which is equal to the product of the field strength vector, $\vec{E}$, (independent of the charged state of the particle) and some scalar quantity called the charge on the drop, q, that is,

$$\vec{F}_E = q\vec{E} \tag{10.1}$$

If e is the elementary unit of charge [in cgs units = 4.8×10^{-10} electrostatic units], then

$$q = ne \tag{10.2}$$

where n is the number of elementary units of charge on the particle. The algebraic sign of the charge is conventionally determined in such a way that the particle is repelled by a charge of a similar sign.

EXAMPLE 10.1

A 10-μm-diameter unit density sphere carries a negative charge equal to 100 electrons. If is placed in an electrical field having a strength of 10 statvolts/cm, determine the force in dynes acting on the particle.

$$\vec{F} = q\vec{E} = ne\vec{E}$$
$$= (100)(4.8 \times 10^{-10})(10)$$
$$F = 4.8 \times 10^{-7} \text{ dynes}$$

This is a very small force, less than one hundred-millionth of the force required to lift a fly.

The cgs electrical units are such that when charge is given in esu and field strength is in statvolts per centimeter, the resulting force is in dynes. The direction of the force is the same as the field except that negatively charged particles will be attracted toward the positive end of the field, and vice versa.

It is customary that electrical parameters be given in terms of "practical" units (volts, amperes, coloumbs, etc.), so conversion factors are required. Practical units are used so that the numbers usually encountered will have values which are not extremely large or small. See Appendix D for a more complete discussion of electrical units.

PARTICLE MOBILITY

The motion of a particle in an electric field depends on two electrical factors, field strength and particle charge. The motion of particles having varying charges and sizes can be compared by considering what their velocities would be in an electric field of unit strength. This velocity, called the *particle mobility*, Z_p, is defined by setting qE equal to 3 $\pi\mu v d$, and solving for v. Then when E equals unity, v becomes the particle mobility Z_p, or

$$Z_p = \frac{qC_c}{3\pi\mu d} \tag{10.3}$$

EXAMPLE 10.2

Determine the mobility of a 10-μm-diameter unit density sphere when it carries 100 unit charges.

$$Z_p = \frac{qC_c}{3\pi\mu d} = \frac{neC_c}{3\pi\mu d}$$

$$Z_p = \frac{(100)(4.8 \times 10^{-10})(1)}{(3)(3.14)(1.83 \times 10^{-4})(10^{-3})}$$

$$Z_p = 0.028 \text{ cm/sec}$$

This represents the velocity the particle would attain when placed in an electrical field having a strength of 1 statvolt/cm.

If the particle mobility is known, it is easy to determine the electrical force acting on the particle, provided field strength is also known. However, field strength may not be constant but may have some spatial or temporal distribution, that is, $\vec{E} = f(x,y,x,t)$. In addition, q may vary from particle to particle and may vary on a single particle with time in a discontinuous, stochastic manner. Thus except for quite simple cases it is exceedingly difficult to predict particle motion in an electric field with accuracy.

Some appreciation of the electrical behavior of aerosols can be gained, however, by considering separately the two factors in Eq. (10.1), q and $\vec{E}$.

PARTICLE CHARGE, q

Particles can be electrified by a number of different sources acting singly or in combination. The basic processes which give rise to a charge on a particle are direct ionization, static electrification, collisions with ions or ion clusters (either with or without an external electric field present), or ionization of the particle by electromagnetic radiation such as ultra violet light, visible light, or gamma radiation. These processes can be considered separately.

Direct Ionization of the Particle

Little is known of this electrification mechanism. For one thing, aerosol densities are generally so small that even though one would expect more ionization taking place in a particle than in an equal volume of air, there are generally at least several orders of magnitude more air mass than particle mass per unit volume of space. Since ionization is primarily a mass dependent phenomenon, there will be at least several orders of magnitude more ionization taking place in the air than in the suspended particles. Thus particle charging should result more from attachment of air ions than by direct ionization. Direct ionization of the particle is not an important particle charging mechanism.

Static Electrification

A second particle charging mechanism is static electrification. This mechanism arises from one or a combination of several other mechanisms, making theoretical interpretation in terms of a single mechanism very difficult, if not impossible (and most experimenters have attempted to interpret their results in terms of a single mechanism). Five basic mechanisms can result in static electrification. These can each be examined for their importance in aerosol physics.

ELECTROLYTIC EFFECTS

In this case solutions of liquids of high dielectric constant exchange ions with metals or solid surfaces. For example, a drop of a high dielectric liquid swept from a metal surface will develop and can carry away a high charge. For a given surface and liquid, droplets will all have a net charge of the same sign, so that the droplets will repel each other. This is probably an important mechanism in aerosol charging, although its importance is not well established. Table 10.1 lists dielectric constants for various materials.

CONTACT ELECTRIFICATION

A second static electrification process is contact electrification. Here electrons migrate from clean, dry surfaces of dissimilar metals to metals with lower work

Table 10.1. Dielectric constants of liquids (esu),
(normal temperature, 20°C)

Oil	2–2.2
Turpentine	2.2–2.3
Methyl alcohol	31
Ethyl alcohol	24.3
Sodium chloride	5.9
Water	78
Magnesium oxide	9.65
Glass	5–10
Polyethylene	2.25
Air	1
CCl_4	2.2
PVC	3.3–4.5

functions. This process requires that there be no impurities between surfaces and is strictly electronic in nature. Because of this requirement, contact electrification is probably not an important charging mechanism for aerosols.

SPRAY ELECTRIFICATION

A third static electrification process is spray electrification. Surface forces in liquids of high dielectric constants increase the concentration of electrons or negative ions in the outer liquid surface (Lenard, 1915). The disruption of these surfaces by atomization or bubbling imparts a predominantly negative charge to the smaller droplets, while the larger ones will be neutral, positive, or negative in approximately equal proportions. The size of all droplets produced may be altered by subsequent evaporation or condensation. Dissolved salts generally reduce the magnitude of the charge compared to charges produced in pure liquids, and the effect is usually reduced as the dielectric constant of the liquid is reduced, until a point is reached as in the case of pure hydrocarbons where little charging is observed. The charged droplets produced by spray electrification generally have only several units of charge per drop. Spray electrification is important in aerosol charging, and very often operates in conjunction with electrolytic effects. This tends to confuse and complicate any attempt at analysis.

TRIBO ELECTRIFICATION

The fourth static electrification method is frictional or tribo electrification. In this mechanism charge is imparted to dry nonmetallic particles when they come in contact with metals or with other particles. Although tribo electrification is a very common charging mechanism, reasons for its occurrence remain fairly obscure. Some points are well known. For example, it is possible to estimate the

Table 10.2. Charge preference in frictional charging

+End
Asbestos
Glass
Galcite
Quartz
Magnesium
Lead
Gypsum
Zinc
Copper
Silver
Silicon
Sulfur
–End

sign of each charge when two different materials come into contact. This is shown in Table 10.2. Materials high on this table will be most likely to develop a positive charge on contact, while those on the lower end are most likely to develop a negative charge. These charges can be produced by particle-particle interaction or by particle-surface interaction, although particle-particle interaction seem to produce more highly charged particles (Miller and Heineman, 1948).

EXAMPLE 10.3

Quartz particles flow through a glass tube. Estimate the sign of the charge produced by static electricity on the particles and the tube. From Table 10.2,

charge on particles   –

charge on tube   +

If the tube were made of copper instead of glass, the signs would be

charge on particles   +

charge on tube   –

Many aerosol experiments have suffered because this relationship has not been clearly understood.

In the case of high concentrations of explosive dusts flowing through an ungrounded duct, sufficient charge may accumulate on the duct to produce a sparking discharge and resulting explosion. This electrification is inhibited when relative humidities exceed 50 or 60%, thought to be due to the formation of a thin moisture layer on the particles. If the moisture contains sufficient dissolved material to make this layer conductive, the charge will not accumulate. This explanation is consistent with the observation that relative humidity, not absolute humidity is important in dust explosions, since deposition of water on the particles, not the presence of water vapor, prevents charging by tribo electrification.

FLAME IONIZATION

A final static electrification method is the ionization of particles in a flame. This effect was first observed as early as 1600, and has recently become the subject of much interest because of potential application in such diverse areas as direct generation of electricity, control of combustion processes by applied electric fields, and the like (Lawton and Weinberg, 1969). In the reaction zones of hydrocarbon/air or hydrocarbon/oxygen flames ion concentrations of 10^9 to 10^{12} ions/cm^3 have been measured. Positive ions are definitely present, but there is some controversy as to whether negative ions or free electrons predominate. The presence of particulate material in the flame (for example, soot particles) greatly enhances the concentration of free charge (Einbinder, 1957). Also, it appears that the smaller the particle size the more free charge that is developed. For example, carbon particles of about 0.02-μm diameter produced in an oxyacetylene flame carried on the average about 10 unit charges per particle, representing an overall charge of 1×10^{18} charges/gram.

Collisions with Ions or Ion Clusters

The best understood of the three main charging mechanisms for aerosols is that involving the collision of ions or ion clusters with aerosol particles. Air ions or ion clusters arise from a number of processes. They can be formed by attachment of either positive or negative charges produced by alpha, beta, or gamma photons as they lose energy following emission from a radioactive source (Cooper and Reist, 1973) or from various types of electrical discharges.

Two distinct processes are involved in charging that can act either singly or in combination. In the first process, *diffusion charging,* particles are charged in the absence of an external electrical field by collisions with diffusing ions. With the second method, *field charging,* particles are charged by ions moving in an orderly direction in an external electric field. The two processes can be considered to be analogous to molecular diffusion and convective diffusions. Charging rates are faster for field charging then for diffusion charging. For very small particles, diffusion charging is important even in the presence of an external field.

To study charging mechanisms theoretically for either diffusion charging or field charging it is necessary to make several assumptions regarding the aerosol. First, the particles are assumed to be spherical. This assumption is reasonable for isometric particles. It is also assumed that the particles are monodisperse. The effect of polydispersity complicates but does not invalidate theory. A third assumption is that there are no interactions between individual particles. Finally, the ion concentration and electric field near each particle are assumed to be uniform. These last two assumptions are essentially true for all natural and industrial aerosols. Except in the most extreme cases theory should be adequate without other modification.

DIFFUSION CHARGING

In diffusion charging, particles are charged by unipolar ions (ions having the same sign) in the absence of an applied electric field. Collisions of ions and particles occur as a result of random thermal motion of the ions, the Brownian motion of the particles being generally neglected.

A simple theory for diffusion charging has been developed by White (1963). He considered that ions diffuse in a gas in accordance with the postulates of kinetic theory, except that when an ion strikes a particle it stays, thus accumulating charge. However, this accumulation of charge on the particle produces an electric field that tends to prevent additional ions from reaching the particle. Thus in White's theory the rate of accumulation of charge on a particle decreases as the charge on the particle increases.

The number of ions striking a spherical particle of diameter d per unit time is

$$\frac{dN}{dt} = \frac{NC}{4}(d^2) = \frac{d^2}{4}NC \tag{10.4}$$

where N is the number of ions near the particle and C is the root mean square velocity of the ions. From kinetic theory the density of ions in a potential field varies according to

$$N = N_0 e^{V/kT} \tag{10.5}$$

in which N_0 is the average ion concentration and V is the potential energy per ion. However, for a particle accumulating charge, the potential energy of an ion of the same charge a distance R from the center of the particle with n charges is

$$V = \frac{-ne^2}{R}$$

At the particle surface the ion concentration is given by

$$N = N_0 \exp\frac{-2ne^2}{dkT} \tag{10.6}$$

From Eq. (10.4), the rate of change of ions per unit time, dN/dt, becomes

$$\frac{dN}{dt} = \frac{\pi}{4}d^2 CN_0 \exp\frac{-2ne^2}{dkT} \tag{10.7}$$

For an initially uncharged particle, integration of Eq. (10.7) gives

$$n = \frac{dkT}{2e^2}\ln\left(1 + \frac{\pi dCN_0 e^2 t}{2kT}\right) \tag{10.8}$$

A characteristic charging time t' can be defined as

$$t' = \frac{2kT}{\pi d C N_0 e^2} \tag{10.9}$$

such that Eq. (10.8) can be written

$$n = \frac{dkT}{2e^2} \ln\left(1 + \frac{t}{t'}\right) \tag{10.10}$$

Furthermore, a characteristic charge, n', can be defined as

$$n' = \frac{dkT}{2e^2} \tag{10.11}$$

so that Eq. (10.8) can be expressed in the dimensionless form

$$\frac{n}{n'} = \ln\left(1 + \frac{t}{t'}\right) \tag{10.12}$$

Figure 10.1 is a plot of Eq. (10.12) showing charge accumulation by diffusion charging as a function of time.

Figure 10.1.   Plot of dimensionless diffusion charging.

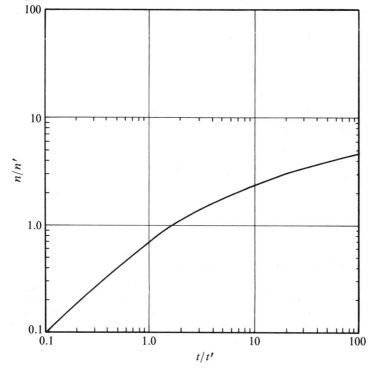

Table 10.3. Representative values of particle charge n for the ion-diffusion charging process

d, cm	t, sec				
	10^{-2}				
10^{-5}	2	3	6	8	10
10^{-4}	35	55	75	95	116
10^{-3}	550	752	954	1,156	1,358
10^{-2}	7522	9542	11,562	13,583	15,603

Table 10.3 gives values for the charge on particles of different sizes at various times, assuming an air temperature of 20°C and $N_0 = 5 \times 10^8$ ions/cm^3. It can be seen from this table, and also from Fig. 10.1, that there is a fairly rapid increase in particle charge initially, followed by a much slower increase later on. No ultimate charge is inherent in this process, however, since the particle is able to charge indefinitely. In actuality the charge on the particle is limited by emission of charge from the particle. It is clear though, that the numerical value of the charge is relatively insensitive to ion concentration and time; whereas it is quite dependent on particle size.

EXAMPLE 10.4

Estimate the charge that would develop in 10 sec by diffusion charging if a 0.5-μm-diameter spherical particle is placed in an ion field containing 5×10^8 ions/cm^3. Assume 20°C temperature.

$$n = \frac{dkT}{2e^2} \ln\left(1 + \frac{\pi d C N_0 e^2 t}{2kT}\right)$$

$$= \frac{(5 \times 100^{-5})(1.38 \times 10^{-16})(293)}{2(4.8 \times 10^{-10})^2} \ln$$

$$\left[1 + \frac{(\pi)(5 \times 10^{-5})(1.18 \times 10^4)(5 \times 10^8)(4.8 \times 10^{-10})^2(10)}{2(1.38 \times 10^{-16})(293)}\right]$$

$$= (4.39) \ln [26,404] = (4.39)(10.18)$$

$$n = 44.67 = 45 \text{ charges}$$

Originally there was criticism that White's derivation only applied to cases where $d > 1$ μm. Liu, Whitby, and Yu (1967a) showed by a more rigorous treatment utilizing the kinetic theory that Eqs. (10.7) and (10.8) are also valid for the case where the ion mean free path is not small compared to the particle size. This derivation was confirmed in experiments carried out by the same three authors (1967b) using corona ions and very small particles. However, to get agreement of the data with theory it was necessary to use a value of 1.18×10^4 cm/sec for the mean thermal speed of the ions, as opposed to a value of 5×10^4 cm/sec for air molecules as used by White. Liu, Whitby, and Yu suggested that this difference could be explained if the ions produced by the corona were associated with molecular clusters, rather than with single molecules. This explanation is consistent with data of Bricard and Pradel (1966).

EXAMPLE 10.5

Estimate the ionic mean thermal speed that corresponds to ions of the hydrated proton $H^+(H_2O)_6$.

From Eq. (3.8)

$$\bar{C} = \sqrt{\frac{8kT}{\pi m}}$$

where m is the ion mass = 109 amu = $\dfrac{109}{6.02 \times 10^{23}}$ g

$$= 1.81 \times 10^{-22} \text{ gr}$$

$$\bar{C}_6 = 23{,}846 \text{ cm/sec}$$

What would this speed be if the hydrated proton was of the form $H^+(H_2O)_{24}$

$$\bar{C}_{24} = \bar{C}_6 \sqrt{\frac{109}{433}} \, \bar{C} = 1.20 \times 10^4 \text{ cm/sec}$$

A more serious fault in White's derivation is in the lack of appreciation of the stochastic nature of the charge acquisition process, as implied by the fairly simple derivation given above. For example, Eq. (10.8) indicates that for small particles and short charging times, fractions of charges are possible. This is clearly an impossibility. Results computed from these equations should be considered to represent average rather than specific values (Boisdron and Brock, 1969; Natanson, 1960).

A more exact derivation which includes the effect of image forces on charge acquisition was developed by Fuchs (1963) and independently by Bricard (1962), and tested experimentally by Liu and Pui (1977).

Figure 10.2 shows a plot of the Fuchs-Bricard theory along with the experimental points of Liu and Pui (1977). Also shown on this figure is a plot of Eq. (10.8), White's equation. Despite the apparent discrepancy in these results, White's equation continues to be used to estimate particle charge acquisition by diffusion charging. This is because it provides a reasonable approximatation of

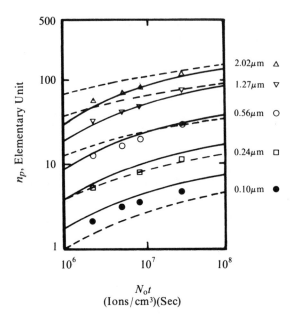

Figure 10.2. Comparison of Fuchs-Bricard theory (solid lines) with that of White (dashed lines). Experimental points of Liu and Pui are also shown.

particle charging and is eminently more simple to use than the more exact Fuchs-Bricard equation.

FIELD CHARGING

Unlike diffusion charging, field charging takes place in a field of unipolar ions, that is, in a region where the ions are in an electric field and hence have ordered motion (Rohmann, 1923; Pauthenier and Moreau-Hanot, 1932). Suppose an uncharged spherical conducting aerosol particle were suddenly placed in a uniform electrical field. The field near the particle would be distorted as illustrated in Fig. 10.3 so that gas ions, following the field lines, would immediately begin

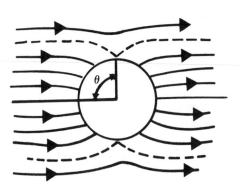

Figure 10.3. Distortion of electric field around an aerosol particle, particle uncharged.

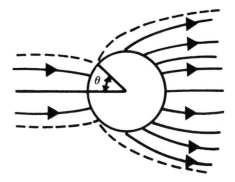

to charge the particle. The dashed lines in the illustration indicate the limits of the field which passes through the sphere. All ions traveling within these limits are considered to strike the particle and charge it.

However, as the particle becomes charged it will start to repel some of the incoming ions. This repulsion results in an alteration of the field configuration which accordingly reduces the charging rate. A point will eventually be reached where no further charging of the particle takes place. This point is known as the saturation charge of the particle. When one-half the saturation charge on the particle is reached, the electrical field surrounding the particle is similar to that shown in Fig. 10.4. Notice that both the ions available to make contact with the sphere and also the particle area available for contact have been reduced.

The ion current to the particle at any time is a function both of the ions that are available to reach the particle, and of the particle area available to accept the ions. Symbolically this is written

$$i = \frac{dq}{dt} = \frac{d(en)}{dt} = jA(n) \tag{10.12}$$

where j is the ion current density in the undistorted field just away from the particle and $A(n)$ is the cross-sectional area of the undisturbed ion stream entering the particle when it is charged with n ions.

The value $A(n)$ is computed from the total electric flux which enters the particle when n ions are present, that is,

$$A(n) = \frac{\psi(n)}{E_0} \tag{10.13}$$

Here $\psi(n)$ is the electric flux entering the particle and E_0 the undistorted electric field strength in the vicinity of the particle.

The electric flux entering the particle is equal to the product of the field at the surface of the particle times the area perpendicular to it, or

$$\psi(n) = \oint E dA \tag{10.14}$$

The electric field E_1 at any point on the surface of a sphere that is placed in an initially uniform electric field can be shown to be

$$E_1 = \zeta E_0 \cos \theta \qquad (10.15a)$$

where $\zeta = 3\epsilon/(\epsilon+2)$ and ϵ is the dielectric constant of the sphere. At the same time, however, charges that have collected on the sphere produce a repelling field that acts to prevent the arrival of additional ions. This repelling field E_2 can be given by

$$E_2 = -\frac{4ne}{d^2} \qquad (10.15b)$$

The net electric field is

$$E = E_1 + E_2 = \zeta E_0 \cos \theta - \frac{4ne}{d^2} \qquad (10.15c)$$

When $\theta = \theta_0$, $E = 0$.

Thus the total electric flux entering the particle is

$$\psi(n) = 2 \int_0^{\theta_0} \left(\zeta E_0 \cos \theta - \frac{4ne}{d^2} \right)\left(\frac{\pi}{2}\right) d^2 \sin \theta d\theta \qquad (10.16)$$

which on integration becomes

$$\psi(n) = \zeta \frac{\pi d^2}{4} E_0 \left(1 - \frac{4ne}{\zeta E_0 d^2} \right)^2 \qquad (10.17)$$

The assumption made here is that the particle is much larger than the ion mean free paths, so that the ions can be considered to follow the lines of force. For large particles (in the continuum region) this assumption is valid. Also, since ion mobility is much greater than particle mobility, partical velocity can be ignored at this point.

The limiting or saturation charge occurs when $\psi(n) = 0$. Setting $\psi(n) = 0$, replacing n with n_s and solving Eq. (10.17) for n_s gives

$$n_s = \frac{\zeta E_0 d^2}{4e} \qquad (10.18)$$

This is the maximum number of charges that can be placed on a particle of diameter d by a field of strength E_0.

EXAMPLE 10.6

Earth's electric field is 1.28 volts/cm over the ocean. What is the maximum electrical charge which can exist on a ten micrometer spherical particle over the ocean due to earth's electric field? Assume $\zeta = 3$.

$$E_0 = 1.28 \text{ volts/cm} = \frac{1.28}{300} = 4.27 \times 10^{-3} \text{ statvolts/cm}$$

$$n_s = \frac{3E_0 d^2}{4e} = \frac{(3)(4.27 \times 10^{-3})(10 \times 10^{-4})^2}{(4.8 \times 10^{-10})}$$

$$n_s = 6.67 \text{ ions/particle; say seven ions}$$

Equation (10.17) can be rewritten in terms of the saturation charge, n_s. Thus

$$\psi(n) = \pi n_s e \left(1 - \frac{n}{n_s}\right)^2 \tag{10.19}$$

The saturation charge represents the maximum charge a particle can reach for a given field strength. If the field strength is made sufficiently intense, a particle will rid itself of excess charge by the spontaneous emission of either electrons or ions. For electrons a surface field intensity of about 10^7 volt/cm is required, while for ion emission a field about 20 times greater is needed (Whitby and Liu, 1966). The number of charges which are implied by these fields thus represents the absolute upper limit on particle charging.

Recalling Eq. (10.12), the second factor to be evaluated is j, the ion current density in the undistorted field. This is the product of the charge per unit volume and the drift velocity of the ions. The charge per unit volume is $N_0 e$ where N_0 is the average ion concentration. When the field energy of the ions is small compared with their thermal energy, the drift velocity of the ions in the field direction is proportional to the electric field intensity, that is,

$$V = ZE_0 \tag{10.20}$$

where Z is called the *mobility* of the ions. For air a typical value for Z is 1.4 cm/sec/volt/cm (McDaniel, 1964). In the cgs system of units this is 420 cm/sec/statvolt/cm. The current density becomes

$$j = N_0 eV = N_0 eZE_0 \tag{10.21}$$

Combining the current density with the total electric flux gives

$$\frac{d(ne)}{dt} = [N_0 eZE_0]\left[\pi n_s e \left(1 - \frac{n}{n_s}\right)^2\right]/E_0 \tag{10.22}$$

or

$$\frac{d(n/n_s)}{dt} = \pi N_0 eZ \left(1 - \frac{n}{n_s}\right)^2 \tag{10.23}$$

which, on integration with the initial condition that $n = 0$ at $t = 0$, gives

$$\frac{n}{n_s} = \frac{\pi N_0 eZt}{\pi N_0 eZt + 1} \tag{10.24}$$

The factor $\pi N_0 eZ$ has the dimensions of the reciprocal of time so a new time factor, t_0, can de denoted as,

$$t_0 = \frac{1}{\pi N_0 eZ} \tag{10.25}$$

and then

$$\frac{n}{n_s} = \frac{t}{t + t_0} \tag{10.26}$$

The factor t_0 can be considered to be a time constant which determines the rate or rapidity of charging; the smaller t_0, the shorter the time of charging to saturation charge. Figure (10.5) shows a plot of Eq. (10.26) indicating that with sufficient time n/n_s reaches the asymptotic value of 1.

Half the final charge is reached at $t = t_0$ and 91% at $t = 10t_0$. Even though larger particles carry much higher saturation charges, the time constant is not size dependent, and relative charging rates of particles of different sizes are the same. Thus in an electrostatic precipitator particles of various sizes placed in the same electric field will charge to the same degree of charge saturation in the same length of time.

Figure 10.5. Plot of fractional saturation charge as a function of dimensionless time. Note that as t approaches infinity, n/n_s approaches 1.

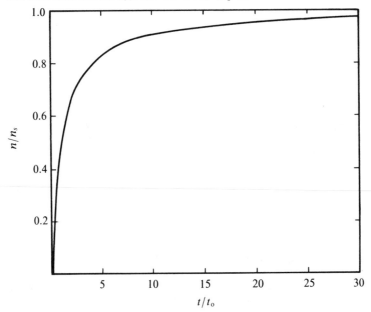

EXAMPLE 10.7

The particle residence time in the charging section of an electrostatic precipitator is 0.4 sec. If the ion concentration is 10^7 ions/cm^3, what fraction of the maximum charge on the particles will be reached in that time?

$$t_0 = \frac{1}{\pi N_0 eZ} = \frac{1}{\pi (10^7)(4.8 \times 10^{-10})(420)}$$

$$t_0 = 0.158$$

$$\frac{n}{n_s} = \frac{t}{t + t_0} = \frac{0.4}{0.4 + 0.158} = \frac{0.4}{0.558} = 0.717$$

That is, approximately 70% of the ultimate particle charge is achieved in 0.4 seconds. Large particles will carry a much greater charge than their smaller counterparts, but the proportions will be the same.

COMBINED DIFFUSION AND FIELD CHARGING

As particle size decreases, charging in an applied electric field results not only from the ordered flow of ions but from random motion of the ions as well. Thus a complete charging theory should account for both phenomena simultaneously. Several difficulties immediately appear. First, diffusion charging places no upper limit on the number of charges a particle may acquire, whereas there is a definite upper limit with field charging. Second, in field charging the particle charge after a given charging period is a function of the square of particle size, while with diffusion charging the charge is approximately a linear function of particle size. With a fairly high applied electric field and small particles the two mechanisms do give comparable results, although when compared with experimental data both mechanisms taken separately tend to slightly underestimate particle charge.

Ion Production by Corona Discharge

The most commonly used method for field charging of aerosol particles is by the use of the phenomenon known as corona discharge. Corona discharge is discussed in detail by White (1963) and Miller and Loeb (1951 b, c) and will only be considered briefly here.

Suppose two electrodes are arranged so that the field strength between them is not constant. (This could be done, for example, with a wire and tube electrode system or a point and plane system.) Then if the potential across the two electrodes is increased, a voltage will be reached where electrical breakdown of the gas occurs nearest either the wire or point. This breakdown is usually manifested by a blue glow, called a corona discharge. With a corona discharge two distinct

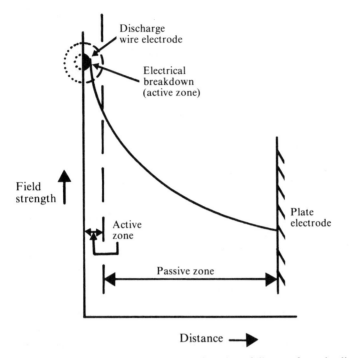

Figure 10.6. Plot of field strength as a function of distance from the discharge wire electrode for a corona discharge.

electrical zones are produced (Fig. 10.6). In the first zone, immediately around the corona wire and containing the corona glow, local electrical breakdown of the gas takes place, caused by collisions with gas molecules of ions leaving the corona wire. If these ions are sufficiently accelerated, the collisions will free additional ions from the molecules. These new ions are also accelerated and, in turn, produce even more ions by collision. Oppositely charged ions are accelerated toward the corona wire where they produce additional ions on impact. This process produces a large number of ions of one sign which rapidly move out of the zone of corona glow toward the other electrode (Fig. 10.7).

As the ions leave the zone of high field strength they tend to attach themselves to gas molecules producing a cloud of slow moving ions all having the same sign of charge as the center electrode, either positive or negative. The corona is said to be negative if a cloud of negative ions is formed, and positive if positive ions are formed. The ions moving toward the passive electrode thus make up the unipolar charging field for aerosol particles. Ion concentrations are typically of the order of 10^7 to 10^9 ions/cm^3. Since electron attachment coefficients and ion mobilities vary greatly from gas to gas, corona characteristics will differ greatly depending on the predominant gas and impurities present. For

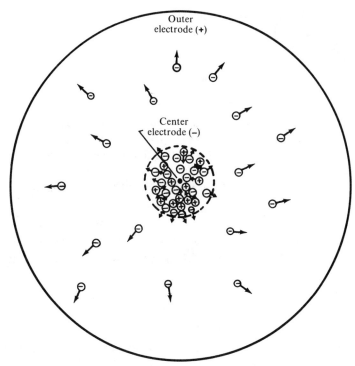

Figure 10.7. Schematic diagram of negative corona discharge showing negative ion motion away from center electrode, positive ion motion toward center electrode.

example, with nitrogen alone negative ion formation is not possible, so that the oxygen component of air is necessary for effective negative particle charging.

With a wire in cylinder arrangement, a negative corona produces tufts or beads of glow along the wire length while a positive corona produces a continuous glow along the wire. Generally a negative corona is preferred for particle charging because it is more stable than the positive corona and can be operated at higher potentials and current flow before sparking occurs, both of which are favorable to electrostatic precipitation. On the other hand, with high electrical potentials, ozone is produced. It appears that a positive corona produces less ozone than a negative one. Thus for cleaning air that will subsequently be supplied to a room or building, a positive corona for particle charging is preferred, since less ozone will be produced, even though the air cleaning efficiency will be somewhat lower.

Charge density in the zone of low field strength depends on ion mobility, which depends on the constituents of the gas being ionized. Nitrogen, hydrogen, and the inert gases absorb few electrons on collision ionization, so charges present in these gases are electrons, having high mobilities and hence a high corona current. Gases such as oxygen, water vapor, sulfur dioxide, and carbon dioxide have a high electron affinity so that negative ions consist almost entirely of gas ions. These gases are called electro-negative gases. The corona current for these gases is relatively low.

Maximum Attainable Particle Charge

In deriving the field charging equation it was shown that for a given field strength and particle size there exists a maximum possible particle charge. It was pointed out that when the field strength reaches the surface field strength for spontaneous emissions of electrons, then the upper limit of particle charging would be established. For a solid spherical particle this limit, n_m, is given by

$$n_m = \frac{E_s d^2}{4e} \tag{10.27}$$

where E_s is the surface field intensity at which emission of ions or electrons occurs. For electrons $E_s \approx 3.3 \times 10^4$ statvolts/cm while for ion emission $E_s \approx 6.67 \times 10^5$ statvolts/cm.

EXAMPLE 10.8

Determine the maximum positive charge on a 0.01-μm-diameter sphere.

$$n_m = \frac{E_s d^2}{4e} = \frac{6.67 \times 10^5 [10^{-6}]^2}{4 \times 4.8 \times 10^{-10}}$$

$$n_m = 347.4 \text{ units of charge.}$$

With a liquid droplet this maximum charge cannot be reached except in the case of extremely small droplet sizes. This is because of an additional charge limitation placed on liquid aerosols, known as the Rayleigh limit. It has been known for many years that as a highly charged droplet evaporates a point will be reached where the outward force of the electric field at the drop surface exceeds the inward force of the droplet's surface tension. At this point the drop will be torn apart by the close proximity of like charges and will produce a number of smaller drops in order to create more surface area for the charge. The number of electrons necessary for droplet disintegration was deduced by Rayleigh as

$$n_r = \frac{1}{e} \sqrt{2\pi\gamma d^3} \tag{10.28}$$

where γ is the surface tension of the liquid. Several experiments have confirmed the validity of this expression (Whitby and Liu, 1966).

Table 10.4   Approximate maximum number of elementary charges on particles

Limit	Particle Diameter, μm		
	0.01	1.0	100
Ion limit	3.47×10^2	3.47×10^6	3.47×10^{10}
Electron limit	1.72×10^1	1.72×10^5	1.72×10^9
Rayleigh limit			
($\gamma = 21$ dynes/cm)	2.39×10^1	2.39×10^4	2.39×10^7
($\gamma = 72.7$ dynes/cm)	4.45×10^1	4.45×10^4	4.45×10^7

EXAMPLE 10.9

Determine the Rayleigh limit for charge on a 1-μm-diameter water droplet ($\gamma = 72.7$ dynes/ cm).

$$n_r = \frac{1}{e} \sqrt{2\pi\gamma d^3}$$

$$= \frac{1}{4.8 \times 10^{-10}} [\pi(72.7)(10^{-4})^3]^{1/2}$$

$$n_r = 4.45 \times 10^4 \text{ units of charge}$$

Table 10.4 lists the approximate maximum number of elementary charges on particles of various sizes for the ion, electron, and Rayleigh limits. For comparison a 1-cm-diameter raindrop in a thunderstorm carries about 4×10^8 charges (Sartor and Atkinson, 1967) or about 1% of its maximum possible charge. Since for all but the smallest particle sizes the Rayleigh limit is the lowest, highly charged drops which can evaporate will disintegrate until drop diameters on the order of 0.01 μm are reached.

EXAMPLE 10.10

Considering both ion and electron limits, find the droplet diameters where the Rayleigh limit just equals these limits. That is, find the droplet diameter which cannot disintegrate upon evaporation.

$$n_{\text{ion}} = \frac{E_s d^2}{4e}$$

$$n_{\text{Rayleigh}} = \frac{1}{e} \sqrt{2\pi\gamma d^3}$$

equating and solving for $d_=$ gives

$$d_= = \frac{32\pi\gamma}{E_s^2}$$

from which the following table can be computed:

	Surface Tension	
	Alcohol $\gamma = 21 \, dynes/cm$	*Water* $\gamma = 72.7 \, dynes/cm$
Electron limit	0.019 μm	0.067 μm
Ion limit	0.00005 μm	0.0002 μm

Positive charge will continue to disintegrate the droplet to molecular size, negative charge will indeed produce a droplet with a finite lower diameter limit.

CHARGE EQUILIBRIUM

In previous sections charging of aerosols by ions of one sign has been discussed. Very often, however, ions of both signs are present in essentially equal numbers. In this case extremely high charges of one sign are not likely to be found on any aerosol particles. However, the presence of free ions suggests that some particles will carry charges. This is particularly true for atmospheric aerosols since there are always free ions available for particle charging. For example, Yaglou and Benjamin (1934) reported an ion range of roughly 200 to 400 small ions/cm^3 of either sign both for indoor and outdoor Boston air, the maximum tending to occur more in the warm summer months. Ion production rates at a meter above the land portion of the earth's surface have been estimated by Wait (1934) to be about 10 ions/cm^3/sec, 2 ions/cm^3/sec coming from cosmic radiation and the remainder from the decay of natural radioactivity emanating from the ground. Since these emanations are not present over oceans, ion concentrations over oceans are much lower than over land. There are essentially the same numbers of positive and negative ions present at any time.

When ions are associated with molecular clusters they are called *small* ions; when attached to small aerosol particles they are often called *large* or Langevin ions (Fleagle and Businger, 1963). The average life of a small ion is roughly 100 sec, that of a large ion about 10-fold longer, or about 1000 sec.

The relatively short lifetime of a charge on an aerosol particle implies charge transfer or neutralization, whereas the continued production of ions suggests a replenishment of the particle charge. Thus if there is an equilibrium value of small ions in the atmosphere there should also be an equilibrium value of charge on aerosol particles present. This equilibrium condition implies that for a given size aerosol particle there should be a definite fraction having no charge, another fraction having one charge, another having two charges, etc. Although any given particle may be gaining or losing charge continually, the aerosol as a whole should maintain the same proportion of charged particles.

Steady-State Theory of Charge Equilibrium

A theoretical approach defining bipolar charge equilibrium has been developed by Keefe, Nolan, and Rich (1959) and comparison with experimental data suggests that it provides a reasonable model for particle sizes from about 0.01 μm to at least 2 μm (Whitby and Peterson, 1965). Keefe, Nolan, and Rich applied Boltzmann's law to the distribution of particle charges in dynamic electrical equilibrium. The usual statement of this law is that the number of particles per unit volume having an energy E, $N(E)$, is given by

$$N(E) = A \exp(-E/kT) \tag{10.29}$$

where A is a normalization constant. In the case of a charged spherical particle carrying n unit charges with a diameter d,

$$E = E_0 + \frac{n^2 e^2}{d} \tag{10.30}$$

Here E_0 represents the energy of the particle in the absence of any charge, whereas the second term represents the additional electrostatic energy. A particle will have the same energy whether it carries a positive or negative charge since the square of the charge is used in Eq. (10.30).

Substituting E given by Eq. (10.30) into Eq. (10.29) gives N_n, the number of particles having n elementary units of charge (of one sign):

$$N_n = N_0 \exp \frac{-n^2 e^2}{dkT} \tag{10.31}$$

for $n > 0$.

The term N_0 represents the number of neutral particles per unit volume given by $N_0 = A \exp(-E_0/kt)$. The number of particles per unit volume carrying n charges of either sign is twice that given in Eq. (10.31), assuming the number of positive and negative particles are equal.

The total number of positively charged particles, N_+, or negatively charged particles, N_-, per unit volume is

$$N_+ = N_- = N_1 + N_2 + N_3 + \cdots + N_\infty \tag{10.32}$$

and the total number of particles per unit volume is

$$N_T = N_0 + N_+ + N_- \tag{10.33}$$

The fraction of particles having n units of charge of one sign, $f(n)$, is

$$f(n) = \frac{N_n}{N_T} = \frac{N_0 \exp(-n^2 e^2/dkT)}{N_0 + \sum_1^\infty 2N_0 \exp(-n^2 e^2/dkT)} \qquad (10.34)$$

or

$$f(n) = \frac{\exp(-n^2 e^2/dkT)}{\sum_{-\infty}^\infty \exp(-n^2 e^2/dkT)} \qquad (10.35)$$

It is interesting to note that the equilibrium charge distribution on aerosols is independent of both ion concentration and aerosol concentration. These factors are important, however, in establishing the length of time necessary for equilibrium conditions to develop.

EXAMPLE 10.11

Determine the fraction of 0.5-μm-diameter aerosol particles (assume spherical shape) at charge equilibrium that carry 2 units of positive charge.

$$\exp(-n^2 e^2/dkT) = \exp \frac{-n^2(4.8 \times 10^{-10})^2}{5 \times 10^{-5}(1.38 \times 10^{-16})(293)}$$

$$= \exp(-0.114 \times n^2)$$

From Eq. (10.34)

$$f(n) = \frac{\exp(-n^2 e^2/dkT)}{1 + \Sigma 2\exp(-n^2 e^2/dkT)}$$

n	$\exp(-n^2 e^2/dkT)$
6	0.017
5	0.058
4	0.161
3	0.359
2	0.634
1	0.892
	2.121

$$f(n) = \frac{0.634}{1 + 2(2.12)} \cong 0.121$$

or about 12% of the particles carry 2 units of positive charge.

Consider the terms in the exponent of Eq. (10.31). With the exception of n and d they are constant for a given temperature. For example, at $20°C$, $e^2/dkT = 5.70 \times 10^{-6}/d$ when d is expressed in centimeters. Denoting this quantity as y gives

$$\frac{N_+}{N_0} = \frac{N_-}{N_0} = e^{-y} + e^{-4y} + e^{-9y} + \cdots \qquad (10.36)$$

With particles greater than 10^{-6} μm in diameter this series can be approximated by

$$\frac{N_+}{N_0} = \frac{N_-}{N_0} = \frac{1}{2}\left(\sqrt{\frac{\pi}{y}} - 1\right) \qquad (10.37)$$

Then the ratio of uncharged particles to total particles becomes

$$\frac{N_0}{N_T} = \frac{N_0}{N_0 + 2N_+} = \sqrt{\frac{y}{\pi}} \qquad (10.38)$$

Very few small particles ($d < 0.1$ μm) will naturally carry *any* charge.

Equation (10.35) can be rewritten utilizing Eq. (10.38), giving

$$f(n) = \sqrt{\frac{e^2}{dkT\pi}} \exp\left(\frac{-n^2 e^2}{dkT}\right) \qquad (10.39)$$

a much more convenient form for computing the equilibrium fraction of charge on various aerosol particles. At temperatures roughly equal to room temperature this equation is applicable to all particles having diameters greater than 10^{-2} μm. For room temperature ($20°C$) this equation becomes

$$f(n) = 0.240 \sqrt{\frac{1}{\pi d}} \exp\left(\frac{-0.058 n^2}{d}\right) \qquad (10.40)$$

when d is expressed in micrometers.

Table 10.5 shows the equilibrium charge distribution on various monodisperse aerosols as computed according to Eq. (10.40).

Table 10.5   Equilibrium charge distribution–Fraction of charges of either sign

| | Number of Charges on Particle | | | | | | | | | |
d	0	1	2	3	4	5	6	7	8	Average Charge
0.01	0.994	0.006								0.006
0.02	0.948	0.052								0.052
0.05	0.606	0.380	0.012							0.403
0.1	0.428	0.479	0.084	0.005						0.662
0.2	0.303	0.453	0.190	0.045	0.006					0.992
0.5	0.191	0.341	0.241	0.135	0.060	0.021	0.006	0.001		1.616
1.0	0.135	0.256	0.215	0.161	0.107	0.064	0.034	0.016	0.007	2.300
2.0	0.096	0.186	0.171	0.148	0.120	0.093	0.067	0.046	0.030	3.251
5.0	0.061	0.120	0.116	0.109	0.101	0.091	0.080	0.069	0.058	5.061

The average number of charges per particle can be determined by adding the charges on all the particles and dividing by the total number of particles. In terms of the fraction of charged particles, $f(n)$, the average number of charges per particle, $\bar{n}$, is

$$\bar{n} = \sum_{-\infty}^{\infty} |n| f(n) \tag{10.41}$$

By replacing the summation with an intregal, Eq. (10.41) becomes

$$\bar{n} \cong \int_{\infty}^{\infty} |n| f(n) \, dn \tag{10.42}$$

which yields, on integration,

$$\bar{n} \cong \sqrt{\frac{dkT}{\pi e^2}} \tag{10.43}$$

a convenient form for determining the average charge for all particles whose diameters are larger than 10^{-1} μm. It should be kept in mind that n represents the average number of charges, *regardless of sign*. The average number of positive or negative charges is half this value.

EXAMPLE 10.12

Determine the average charge per particle for an aerosol made up of 0.5-μm-diameter spheres.

For these particles,

$$\bar{n} = \sqrt{\frac{dkT}{\pi e^2}}$$

$$\bar{n} = \sqrt{\frac{(5 \times 10^{-5})(1.38 \times 10^{-16})(293)}{\pi(4.8 \times 10^{-10})^2}}$$

$$\bar{n} = 1.67 \text{ charges/particle}$$

Transient Approach to Charge Equilibrium

Experimental data indicate that given enough time and otherwise optimum conditions, an equilibrium charge will eventually develop on an aerosol. Very often, however, it is of interest to know whether this charge distribution has in fact developed, and to gain insight into the factors which could be changed to hasten or retard its development. Exact calculation of the transient approach to charge equilibrium is extremely difficult. It is more appropriate to use an equilibrium half-time, similar to the half-life in radioactive decay, to describe the rate at which charge equilibrium is being reached. This represents the length of time necessary for half the equilibrium charge to be attained, and is (Flanagan and O'Connor, 1961)

$$t_{1/2} = \frac{0.693 \, N_T}{4 \, q_i} \qquad (10.44)$$

where q_i is the ion production rate. This equation indicates that with increased ion production rates or decreased aerosol concentration, charge equilibrium is more quickly reached, a fact borne out by experiment.

With an ion production rate of 10^4 ions/cm³-sec, and an aerosol concentration of 5×10^4 particles/cm³, equilibrium would be achieved in about 2 seconds. For atmospheric aerosols where the ion production rate may be only 10 ions/cm³-sec, even though aerosol concentrations of 5×10^4 particles/cm³ are not uncommon, it takes approximately 1700 seconds (or about 30 minutes) for equilibrium to be achieved. O'Connor and Sharkey (1960) report that equilibrium conditions usually prevail in air coming from the ocean. Over an industrial city, on the other hand, measurements indicated that the equilibrium charge distribution was not attained (Nolan and Doherty, 1950). This difference was attributed to the shorter time span between the production of the aerosol over the city and its measurement.

Since charge equilibrium can be quickly attained by using high ion production rates or large ion concentrations, it is not surprising to find this method employed for aerosol charge neutralization. Here the idea is to use the large number of free ions to reduce the excess charge on highly charged aerosol particles to as low a value as possible. With a mixture of bipolar ions, charge equilibrium as discussed in the previous sections will be rapidly attained. This method was developed by Whitby (1961) and Whitby and Peterson (1965), and has been subsequently applied with great success.

Radioactive sources can also be used for charge neutralization, since these produce large numbers of bipolar ions that can then rapidly neutralize highly charged aerosols (Cooper and Reist, 1973).

PROBLEMS

1. A 5-μm-diameter unit density sphere carries a negative charge equal to 200 electrons. If it is placed in an electrical field having a strength of 1000 volts/cm, determine the force in dynes acting on the particle.

2. Determine the mobility of a 5-μm-diameter unit density sphere when it carries 200 unit charges.

3. Lead particles flow through a rubber tube. Estimate the sign of the charge produced by static electricity on the particles and the tube.

4. Estimate the charge which would develop in 60 sec by diffusion charging if a 0.25-μm-diameter spherical particle is placed in an ion field containing 3×10^8 ions per cm³. Assume $20°$C temperature.

5. The particle residence time in the charging section of an electrostatic precipitator is 0.6 sec. What is the ion concentration such that one-half the maximum charge on the particles is reached during this residence time?

6. What fraction of 0.5-μm particles will have an average of three charges on
them at equilibrium? What fraction will have an average of four charges?

7. Plot a curve showing charge as a function of time for diffusion charging using
the terms $2\,e^2 n/dkT$ on the y-axis and the corresponding dimensionless term
on the x-axis.

8. Plot a curve for field charging of a 1-μm sphere showing fraction of total
charge as a function of dimensionless time.

ELEVEN

ELECTROSTATIC CONTROLLED
AEROSOL KINETICS

ELECTRIC FIELDS

As pointed out in Chapter 10, not only must the number of electric charges carried by the particle be known to determine the electrical force acting on an aerosol particle, but also the strength of the electric field acting on these charges. Electric field strength is a vector quantity having both magnitude and direction. The strength of the field is indicated by the number of lines of force passing through each unit area of orthogonal surface. As an example, Fig. 11.1 shows field direction lines (solid lines) and lines of force (dotted for several different geometries). The number of lines of force through a unit area is called the flux or induction through that area.

FIELD STRENGTH OF A POINT CHARGE

The flux through an arbitrarily oriented element of area, *ds,* can be shown to be

$$d\phi = \vec{E}\, ds \qquad (11.1)$$

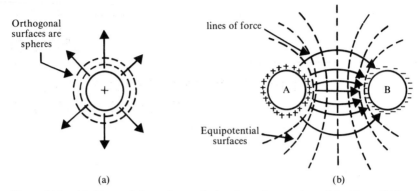

Figure 11.1. (a) Lines of force for a singly charged aerosol particle. Since field strength is force per unit area, field strength decreases as the square of the distance from the particle in this case. (b) Lines of force for two aerosol particles; one with a positive charge, one with a negative charge. Orthogonal surfaces are no longer spheres.

so that the flux through any finite surface s is

$$\phi = \oint \vec{E}\, ds \tag{11.2}$$

When this integral is taken over a closed surface, there may be an excess of lines of force leaving the enclosed volume compared to the number entering. This indicates that a field is originating from within the closed volume. If the integral is taken around an aerosol particle having a charge q, then

$$q \propto \oint \vec{E}\, ds \tag{11.3}$$

indicating that the electric charge on a particle plays a double role. Besides being the object on which an electric field acts, it is also active as the generator of an electric field. This point is important in practical considerations of electrostatic precipitation.

When the particle is represented by a single point charge (Fig. 11.1) the lines of force are radial and equal in all directions. The orthogonal surfaces of equal field strength are spherical surfaces with a common center at the center of the particle, with the flux through any of these spheres of radius r being

$$\phi = \oint \vec{E}\, ds \tag{11.4}$$

Since the density of the lines of force is the same everywhere, the field strength E must be constant over the surface of the sphere so that

$$\phi = \oint \vec{E}\, ds = 4\pi r^2 \, |E| = \gamma q \tag{11.5}$$

Hence, the field strength for a point charge a distance r away from the charge is

$$|E| = \frac{\gamma q}{4\pi r^2} \tag{11.6}$$

where γ is a factor of proportionality.

COULOMB'S LAW

Suppose a second particle of charge q is situated a distance R away from the first particle. Then the force acting on the second particle because of the field generated by the first would be, from Eq. (10.1),

$$F = q'E = \frac{q'\gamma q}{4\pi R^2} \tag{11.7}$$

This is Coulomb's law. The units for charge, field strength, and force are made compatible by specifying the units of the factor of proportionality, γ. For example, if $\gamma = 4\pi/\epsilon$, where ϵ is the dielectric constant of the medium, the units are in terms of cgs or absolute electrostatic system (esu). Since the dielectric constant for air is essentially one, for aerosols using the cgs system of units, $\gamma = 4\pi$.

EXAMPLE 11.1

Two 0.1-μm-diameter unit density spheres, each carrying one positive charge, are situated in air a distance 1 cm apart. Estimate the repelling force between these two particles.

From Eq. (11.7), recalling that $q = ne$,

$$F = \frac{\gamma e\, e}{4\pi R^2}$$

With cgs units

$$F = \frac{e^2}{\epsilon R^2}$$

for air $\epsilon = 1$

$$F = \frac{(4.8 \times 10^{-10})^2}{(1)^2} = 2.3 \times 10^{-19} \text{ dynes}$$

Electrical forces between particles are negligibly small until the particles are almost touching. For comparison, the gravitational force on these particles is almost six orders of magnitude larger than this result. Hence interparticle electrical forces can generally be neglected in aerosol computations.

ELECTRICAL UNITS

Very often "absolute" units, rather than electrostatic or cgs units, are used in dealing with electrical quantities. This is done to do away with the very small and large numbers which occur with cgs units. For the absolute system, γ is defined as

$$\gamma = \frac{1}{K_0 \epsilon} \tag{11.8}$$

The constant K_0 has a value of 8.859×10^{-12} amp-sec/meter-volt. Table D.1 (see Appendix D) lists conversion factors for various electrical parameters to convert from absolute to electrostatic units. For example, the unit charge of an electron, 4.8×10^{-10} statcoulombs (esu) in cgs units becomes, in absolute units, 1.59×10^{-19} coulombs.

GENERAL EQUATIONS FOR FIELD STRENGTH

The electric field strength at any point is the spatial derivative or gradient of the electrostatic potential at that point. The electrostatic potential for various geometries and boundary conditions for regions with no charge is given by LaPlace's equation,

$$\nabla^2 V = 0 \tag{11.9}$$

or for regions having a charge density,

$$\nabla^2 V = -\gamma \rho_s \tag{11.10}$$

known as Poisson's equation, where ρ_s is the space charge per unit volume. The symbol ∇^2 represents the LaPlacian operator (Chapter 8). The problem of calculating the electrostatic field strength is solved by first finding the distribution of potential within the field. Then the derivative of this solution with respect to distance gives the field strength, that is

$$E = -\text{grad } V \tag{11.11}$$

EXAMPLE 11.2

Write an equation for the electrostatic potential that would exist within a wire and tube type electrostatic precipitator (see Fig. 11.2) for

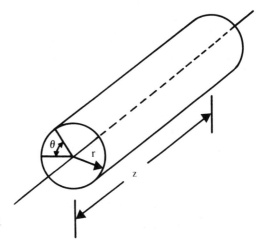

Figure 11.2. Schematic of wire and tube-type electrostatic precipitator.

a. no charge density, and

b. charge density.

a. Since a wire and tube type precipitator is cylindrical in shape, choice of a cylindrical coordinate system is appropriate. Then

$$\nabla^2 V = 0 = \frac{\partial^2 V}{\partial r^2} + \frac{1}{r}\frac{\partial V}{\partial r} + \frac{1}{r^2}\frac{\partial^2 V}{\partial \theta^2} + \frac{\partial^2 V}{\partial z^2}$$

It is not expected that potential will vary with the tube length. Hence $\partial^2 V/\partial z^2 = 0$. Also, potential will not vary with θ. Thus $\partial^2 V/\partial \theta^2 = 0$ and since only one independent variable remains, the equation becomes

$$\frac{d^2 V}{dr^2} + \frac{1}{r}\frac{dV}{dr} = 0$$

b. Here the solution is the same except that a space charge is now present,

$$\frac{d^2 V}{dr^2} + \frac{1}{r}\frac{dV}{dr} = -\gamma\rho_s$$

These equations can be expressed in terms of the field strength by applying Eq. (11.11). Then

$$\frac{dE}{dr} + \frac{1}{r}E = 0 \qquad\qquad (11.11a)$$

and

$$\frac{dE}{dr} + \frac{1}{r}E = \gamma\rho_s \qquad\qquad (11.11b)$$

CONSTANT FIELD STRENGTH

The field strength between two parallel plates plates is a constant, being equal to the potential difference across the plates divided by distance between them. Near the surface of the earth an essentially constant field exists between the negative earth and the positive ionosphere, with field strengths ranging in the order of 0.67 to 3.17 volts/cm over land and about 1.3 volts/cm over sea (Mason, 1971; Pruppacher and Klett, 1978).

COMPUTATION OF THE ELECTRIC FIELD
FOR SIMPLE GEOMETRIES

Very often the electric field is not constant (as in the case of parallel plates) but is spatially dependent. Then it is necessary to determine the field strength as a function of some characteristic distance. Consider a cylinder of radius R having a fine wire running down its axis. This could be a tube, for example, through which an aerosol is flowing. A potential V is established across the wire-tube geometry.

Negligible Ionic Space Charge

When the charge on the center wire is relatively low, the ionic space charge density is assumed to be negligible, and LaPlace's equation is applicable. In cylindrical coordinates assuming cylindrical symmetry, with the axis along the axis of the two cylinders, LaPlace's equation can be written as:

$$\frac{dE}{dr} + \frac{E}{r} = 0 \tag{11.12}$$

(See Ex. 11.2.) Integration gives

$$E = \frac{C}{r} \tag{11.13}$$

where the constant C has the value

$$C = \frac{V}{\ln(r_o/r_i)} \tag{11.14}$$

and r_i is the radius of the inner electrode, r_o the radius of the outer electrode, and V the potential across the electrodes.

EXAMPLE 11.3

An electrostatic precipitator sampler consists of a 0.020-in. diameter wire placed along the axis of a 1.5-in.-diameter tube. What is the maximum field strength (assuming negligible space charge) at the outer edge of the tube when the precipitator voltage is 20 kV.

From Eq. (11.13)

$$E = \frac{C}{r} = \left(\frac{1}{r_o}\right) \frac{V}{ln(r_o/r_i)}$$

$$= \frac{\dfrac{20 \times 10^3}{300}}{\left(\dfrac{1.5 \times 2.54}{2}\right) ln \left[\dfrac{1.5 \times 1/2}{0.02 \times 1/2}\right]}$$

$$E = 8.11 \text{ statvolts/cm.}$$

In practical units the field strength would be about 2430 volts/cm. Under the conditions of the problem, a corona discharge is likely to be found around the center wire, so that the assumption of negligible space charge will not be met. However, the example does illustrate the calculation.

Ionic Space Charge Present

Now suppose the center wire charge is considered, and it is sufficient to produce a corona discharge. This corona, or more exactly, the resulting ions produced, give rise to an ionic space charge within the outer cylinder. Assuming that the wire acts only as an ion source, then the current applied to the wire will be used to maintain this space charge, or ionic current, which can be given as

$$i = 2 \pi r \rho_s ZE \tag{11.15}$$

where Z is the ionic mobility and ρ_s the ion density.

Using Poisson's equation with cylindrical coordinates gives (from Ex. 11.2)

$$\frac{dE}{dr} + \frac{E}{r} - \frac{2i}{ZrE} = 0 \tag{11.16}$$

since

$$\gamma \rho_s = \frac{2i}{rZE}$$

Integrating gives

$$E = \left(\frac{2i}{Z} + \frac{C^2}{r^2}\right)^{1/2}$$ (11.17)

The constant C depends on corona voltage and current, as well as on inner and outer cylinder diameters. For large values of i and r, Eq. (11.17) reduces to

$$E = \sqrt{\frac{2i}{Z}}$$ (11.18)

implying a constant field strength over most of the cross section away from the inner electrode.

An approximation for the corona current, i, has been given by White (1963) to be

$$i = V(V - V_o) \frac{2 \times Z}{r_o^2 \ln[r_o/r_i]}$$ (11.19)

Equation (11.19) represents a reasonably good approximation for relatively low corona currents when V, the operating voltage, is slightly above the corona starting point. The corona starting voltage can be estimated from the expression

$$V_o = 100\delta f r_i \left(1 + \frac{0.3}{\sqrt{r_i}}\right) \ln\left[\frac{r_o}{r_i}\right]$$ (11.20)

where δ is a correction factor for temperature and pressure,

$$\delta = \frac{293}{T} \times \frac{P}{760}$$ (11.21)

Temperature is expressed in degrees Kelvin and pressure in millimeters of mercury. The factor f is a wire roughness factor, equal to unity for a perfectly smooth round wire, but usually in practice having a value lying somewhere between 0.5 and 0.7 (White, 1963).

EXAMPLE 11.4

Determine the field strength for the sampler in Ex. 11.3 considering ionic space charge. The precipitator voltage is 20 kV. Assume 20°C, standard pressure, $f = 0.6$. Use $Z = 2.2$ cm/sec/volt/cm.

From Ex. 11.3, $r_i = 0.02 \times 2.54 \times 1/2 = 0.025$ cm, $r_o = 1.5 \times 2.54 \times 1/2 = 1.905$ cm.

The corona starting voltage is

$$V_o = 100\delta f r_i \left(1 + \frac{0.3}{\sqrt{r_i}}\right) \ln\left[\frac{r_o}{r_i}\right]$$

$$\delta = 1$$

$$V_o = 100(1)(0.6)(0.025) \left(1 + \frac{0.3}{\sqrt{0.025}}\right) \ln \frac{1.905}{0.025}$$

$$V_o = (1.52)(2.88)(4.32)$$

$$V_o = 18.97 \text{ statvolts}$$

$$i = V(V - V_0) \frac{2Z}{r_o^2 \ln [r_o/r_i]}$$

$$= \frac{20,000}{300} \left(\frac{20,000}{300} - 18.97\right) \frac{(2)(2.2 \times 300)}{(1.905)^2 \ln\left[\dfrac{1.905}{0.025}\right]}$$

$$= (66.67)(47.70)(84.25)$$

$$= 2.68 \times 10^5 \text{ statamps/cm}$$

Then

$$E = \sqrt{\frac{2i}{Z}} = \sqrt{\frac{2 \times 2.67 \times 10^5}{660}} = 28.49 \text{ statvolts/cm}$$

Electric Field — Particles Present

Finally, consider the case when there are particles present in the electric field. How is the the field modified by the particle space charge? By neglecting the ion space charge as compared to the particle space charge, White (1963) showed by solution of Poisson's equation that the corona starting voltage, V_o, would be increased by an amount equal to $\pi \rho_o r_o^2$. Then, if V_o' is the corona starting voltage when particles are present,

$$V_o' = V_o + \pi \rho_o r_o^2 \tag{11.22}$$

For N_T spherical particles of diameter d per cubic centimeter carrying the saturation charge,

$$\rho_o = N_T n_s e \tag{11.23}$$

Since $m = N_T(\pi/6)d^3\rho$ where m is the particle mass per cubic centimeter, ρ the particle density, and $n_s e = \zeta E_o d^2/4$, Eq. (11.23) is equivalent to

$$\rho_o = \frac{3\zeta}{2} \frac{mE_o}{\pi d\rho} \qquad (11.24)$$

EXAMPLE 11.5

How much will the corona starting voltage increase in the precipitator of Ex. 11.4 when fly ash particles having an average diameter of 0.1 μm are present if they have been fully charged in a 5 kV/cm field (assume $\zeta = 3$, $\rho_p = 1$ g/cm^3). The particle mass concentration is 0.5 g/M^3.

$$\rho_o = \frac{9}{2} \frac{mE_o}{\pi d\rho}$$

$$\rho_o = \frac{(9)(0.5 \times 10^{-6})(5/0.3)}{2(\pi)(10^{-5})(1)}$$

$$\rho_o = 1.19 \text{ esu/cm}^3$$

$$\pi\rho_o R_o{}^2 = (\pi)(1.19)\left(1.5 \times \frac{2.54}{2}\right)^2$$

increase = 13.6 statvolts = 4080 volts

The effect of the particle space charge is to reduce corona current by increasing the corona starting voltage. Increases in aerosol mass concentrations will increase the effective corona starting voltage, as will decreases in aerosol particle size for a given mass concentration. Thus very fine fumes in high concentration can be quite difficult to remove by electrostatic precipitation.

A second space charge effect is the mutual repulsion by particles carrying charges of similar sign. The effect results in an apparent increase in field strength near the collecting surface which can be approximated by the factor

$$\sqrt{1 + \frac{3mR_o}{d\rho}}$$

so that the field strength becomes

$$E = \sqrt{\frac{2i}{Z}} \times \sqrt{1 + \frac{3mR_o}{d\rho}} \qquad (11.25)$$

In general, this increase in field strength does not offset the reduction in corona current. Hence the net effect of small particles in an electrostatic precipitator is a lowering of collection efficiency.

EXAMPLE 11.6

Compute the field strength for the precipitator in Ex. 11.4 when the fly ash of Ex. 11.5 is included in the calculations.

Corona starting voltage, V_0 = 18.97 + 13.6, from Ex. (11.4) and Ex. (11.5):

$$V_0 = 32.57 \text{ statvolts}$$

$$V = 20 \text{ kV} = \frac{20}{0.3} = 66.67 \text{ statvolts}$$

$$i = V(V - V_o) \frac{2Z}{r_o^2 \ln [r_o/r_i]}$$

$$= 66.67 (66.67 - 32.57) \frac{(2)(660)}{(1.905)^2 \ln \left[\dfrac{1.905}{0.025}\right]}$$

$$= 1.91 \times 10^5 \text{ statamps/cm}$$

$$E = \sqrt{\frac{2i}{Z}} \times \sqrt{1 + \frac{3mR_o}{d\rho_p}} = \sqrt{\frac{3.82 \times 10^5}{660}}$$

$$\times \sqrt{1 + \frac{(3)(5 \times 10^{-7})(1.91)}{0.1 \times 10^{-4}(1)}}$$

$$= \sqrt{579} \times \sqrt{1.29} = 27.3 \text{ statvolts/cm}$$

PERTURBATIONS IN THE ELECTRIC FIELD CAUSED BY A PARTICLE OR OTHER OBJECT

Up to now only coulombic force has been considered. This is the force between a particle and collecting surface due to the net charge on each surface and assuming that the charge on each surface is constant and stationary. Additional electrical forces can also be present. Consider two conducting spherical particles, one with a net positive charge and the other with no net charge. As the first particle approaches the second, the positive charge attracts electrons from the back side of the second to its front (Fig. 11.3), forming a dipole with a net negative charge nearest the oncoming particle. This net negative charge sets up an attracting force between the two particles. This force, which can also arise between a charged particle and uncharged collecting surface or vice versa, is known as a polarization, induction, or image force. It generally tends to enhance the collection of charged particles by *any* surface. It also enhances the collection of uncharged particles by a surface placed in an electric field, although the enhancement is poor for very small uncharged particles since this force is proportional to the volume of each particle.

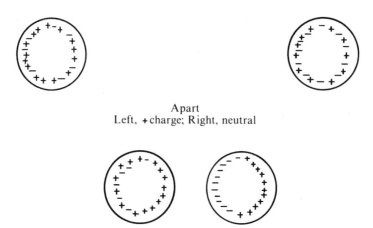

Apart
Left, +charge; Right, neutral

Together
Dipole

Figure 11.3. Mechanism of charge alignment for conducting aerosol particles as they are brought together by an external force.

In many cases it is valid to neglect all electrical forces acting on an aerosol with the exception of the coulombic force. This greatly simplifies most problems, but, if not used with care, can produce significant errors or lead one to erroneous conclusions. For example, Fig. 11.4 shows the trajectories of small positively charged particles in an electric field as they flow around an uncharged fiber; in (a) where the electric field tends to move the particles along with the air flow and in (b), where the field imparts a force on the particles in an opposite direction to the airstream. In the former case deposition can take place, in the latter example it does not. Neglect of image forces would have these two cases equivalent, with deposition occurring only through aerodynamic forces (Hochrainer et al., 1969).

PARTICLE DRIFT IN AN ELECTRIC FIELD

The main reason for evaluating the charges on aerosol particles and the electric fields that act on these charges is to develop models which describe the effect on particle motion of the electrical force.

The equation of motion for an aerosol particle including an electrical force present, F_E, can be written as

$$m \frac{dv}{dt} = F_D + F_G + F_E \tag{11.26}$$

which becomes

$$\tau\frac{d\vec{v}}{dt} = (\vec{u} - \vec{v}) + \tau g\vec{G} - \vec{E}qB \tag{11.27}$$

where B is the particle mobility. When $\vec{u}$ is a constant, $\vec{u}_0$ is the sum of the constant vectors, $(\vec{u} + \tau g\vec{G})$, and Eq. (11.27) becomes

$$\tau\frac{d\vec{v}}{dt} + \vec{v} = \vec{u}_0 - \vec{E}qB \tag{11.28}$$

In terms of a dimensionless velocity, $v' = v/u_0$, we can write, for Eq. (11.28)

$$\tau\frac{dv'}{dt} + v' = 1 - \Gamma \tag{11.29}$$

where the dimensionless parameter, Γ, which can be either positive or negative, is equal to EqB/u_0 and indicates the ratio of the particle velocity in an electric field to the constant velocity u_0. If $|\Gamma| \gg 1$, then electrical forces predominate, whereas when $|\Gamma| \ll 1$, gravity and inertial effects predominate and electrical forces can be neglected. Since in general τ is quite small, when $|\Gamma| \gg 1$ the inertia term $\tau(dv'/dt)$ can be ignored and Eq. (11.29) can be written simply as

$$\vec{v}' \simeq -\Gamma \tag{11.30a}$$

or

$$v \simeq -EqB \tag{11.30b}$$

Figure 11.4.   (a) Field in direction of particle motion. (b) Field in opposite direction of particle motion.

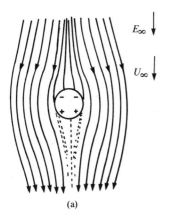 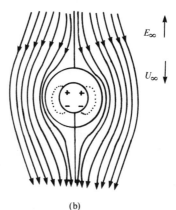

(a)   (b)

Consider the case where $|\Gamma| \gg 1$. The electrical drift velocity is given by Eq. (11.30b). Denoting the particle velocity in an electric field as w and assuming a saturation charge, for field charging Eq. (11.30b) becomes

$$w = E \left(\frac{\varsigma E_o d^2}{4} \right) \frac{C_c}{3\pi\eta d}$$

$$w = \frac{\varsigma E E_o d C_c}{12\pi\eta} \tag{11.31}$$

The term E_o is the field generating the particle charge and E is the collecting field strength.

Example 11.7

Determine the electrical drift velocity of the fly ash particles in previous examples.

From Ex. (11.6), $E = 24.40$ statvolts/cm

From Ex. (11.5), $E_o = 5$ kV/cm $= 16.67$ statvolts/cm

$\varsigma = 3, d = 10^{-5}$ cm, $C_c = 2.78$

$$w = \frac{E E_o d C_c}{12\pi\eta}$$

$$w = \frac{(3)(24.40)(16.67)(10^{-5})(2.78)}{(12)(\pi)(1.83 \times 10^{-4})}$$

$$w = 4.92 \text{ cm/sec}$$

EFFICIENCY OF AN ELECTROSTATIC PRECIPITATOR

The utility of the concept of aerosol particle electrical drift velocity can be shown by using it to estimate the theoretical efficiency of an electrostatic precipitator. For simplicity it is assumed that the collector is cylindrical in shape, having a radius R (although this assumption does not affect the results), and that an aerosol is uniformly distributed across the entrance of the collector. In addition, turbulent flow in the collector is assumed such that the uncollected aerosol remains uniformly distributed at any distance from the entrance of the tube. If the electrical drift velocity is constant, the chance of a particle, p, being collected in a time Δt is

$$p = \frac{w(2\pi R)}{\pi R^2} \Delta t = \frac{2w}{R} \Delta t \tag{11.32}$$

and the chance of not being collected is $(1 - p)$.

In n intervals of time the chance of not being collected is $(1 - p)^n$. When n is allowed to approach infinity during a time period t, $(1 - p)^n$ approaches a value of $\exp(-pt)$. Denoting ϵ as the collection efficiency,

$$\epsilon = 1 - \exp(-2w/R)t \qquad (11.33)$$

In terms of the volumetric gas flow through the tube, Q, the efficiency is

$$\epsilon = 1 - \exp(-Aw/Q) \qquad (11.34)$$

where A is the total collecting area of the precipitator. Equation (11.34) is applicable to both tube and plate type precipitators. It is known as the Deutsch (1922) equation, and its general form has been verified many times in practice.

EXAMPLE 11.8

If the precipitator of Ex. (11.7) is 6 in. long and air flows through it at a rate of 1 cfm, determine the efficiency of collection of this unit for 1-μm fly ash particles.

$$\epsilon = 1 - \exp(-Aw/Q)$$

$$Q = 1 \text{ cfm} = 1 \times 28.3 \text{ l/min} = 472 \text{ cm}^3/\text{sec}$$

$$A = (2)(3.14)\left(\frac{1.5}{2} \times 2.54\right)(6 \times 2.54)$$

$$A = 182.4 \text{ cm}^2$$

$$\epsilon = 1 - \exp\left(-\frac{182 \times 4.92}{472}\right)$$

$$\epsilon = 1 - 0.15 = 0.85 = 85\% \text{ efficient.}$$

It should be kept in mind that the derivation of Eqs. (11.33) and (11.34) contains many simplifying assumptions which may or may not be valid, depending on aerosol and precipitator characteristics. For example, it is assumed that once a particle is collected, it remains collected. This is not the case except for a liquid aerosol. Also, for dry aerosols, when the particles are good conductors they rapidly lose their charge to the collecting electrode and pick up a new charge of opposite sign from the electrode causing them to be repelled. On the other hand, if the particles are poor conductors, they will lose their charges so slowly that the rain of new charges should be sufficient to maintain the charge on the particles and hold them to the collecting surface.

Theoretical calculations will always overestimate precipitator efficiencies, probably because of reentrainment. This overestimation could be as large as a factor of 2 or more (Rose and Wood, 1966). Even so, drift velocity or "effective migration velocity" is the basis for all precipitator calculations and does provide a good base for the comparison of various designs.

PROBLEMS

1. A 0.1-μm-diameter unit density sphere and a 0.2-μ-diameter unit density sphere, each carrying two positive units of charge, are spaced in air a distance 1 cm apart. Estimate the repelling force between these two particles.

2. An electrostatic precipitator sampler consists of a 0.015-in.-diameter wire placed along the axis of a 1-in.-diameter tube. What is the maximum field strength (assuming negligible space charge) at the outer edge of the tube when the precipitator voltage is 15 kV.

3. Determine the field strength for the sampler in problem 3 considering ionic space charge. The precipitator voltage is 15 kV. Assume 20°C, standard pressure, $f = 0.6$. Use $Z = 2.2$ cm/sec/volt/cm.

4. Determine the electrical drift velocity of 0.1-μm-diameter spheres having a density of 2.65 g/cm^3 if they are carrying 200 units of charge each and are placed in a collecting field of 70,000 volts/meter.

5. An electrostatic precipitator is to be used to control emission of 0.5-μm-diameter particles from a paper mill. An efficiency for the collector of 99.6% is desired. If the total design flow through the unit is to be 6500 cfm, how many square feet of collector surface are required? Assume $w = 7.5$ cm/sec.

CONDENSATION AND EVAPORATION PHENOMENA IN AEROSOLS

Condensation and evaporation of aerosols play a great part in man's existence. The cycle of water in nature relies on the condensation of water to form cloud droplets, some of which then return to earth in the form of rain or snow. Photographs of the earth's surface taken from outer space reveal that the most distinguishing characteristic of the earth is its cloud cover. Clouds and fogs lower visibility and can have a marked effect on air temperatures at the earth's surface. Fogs in combination with man-made air pollution can result in aerosols which are quite irritating to man as well as being toxic to some forms of plant life (and, in some cases, to human life as well). Many industrial pollutants appear as aerosols made up of condensed liquids.

Evaporation of liquid drops is equally important. For example, in the application of a pesticide by spraying it is desired that evaporation be minimized to increase the amount of pesticide reaching the plants. On the other hand, in the production of such foodstuffs as powdered milk or powdered coffee, product quality is improved when evaporation proceeds as quickly as possible. In sampling aerosols, evaporation or condensation may alter aerosol size distribution and affect operation of the sampling instrument. In this case it is desired that static conditions be maintained if at all possible.

EARLY OBSERVATIONS

Early investigators such as Coulier (1875) and Aitken (1880) found that when they produced clouds by the adiabatic expansion of moist air (no heat transfer between the system and surrounding container) the presence of small dust particles was necessary for cloud formation. If the air were first made dust-free, clouds would not form. In this case clouds only appeared when the expansion was very large. C. T. R. Wilson (1897) extended these studies by defining the conditions under which clouds could be formed without dust particles: spontaneously with very high supersaturations or at lower supersaturations when ions were present. It was these observations that led to the development of cloud chambers for ion track visualization.

TYPES OF NUCLEATION

Early investigators determined that the formation of an aerosol initially required a surface for condensation. This surface could be made up of a small cluster of vapor molecules, an ion or ionic cluster, or it could be a small particle of some other material, termed a condensation nucleus. When condensation of a vapor takes place solely on clusters of similar vapor molecules, it is called *spontaneous* or *homogeneous nucleation*. When condensation occurs on a nucleus or dissimilar material, it is called *heterogeneous nucleation*.

In the case of homogeneous nucleation, supercooling of the liquid making up the drop is common when the drop temperature is lowered below the freezing point, since there are no foreign bodies present in the liquid. For water droplets, supercooling to temperatures as low $-40°C$ is possible. With a single condensation nucleus in the drop, its purity is such that supercooling is still quite common. This implies that in the formation of any particle by condensation (solid or liquid) it goes through a liquid phase (although the time the particle remains in this phase might be very short) and thus the theory developed for condensation and evaporation of liquid aerosols can also be applied to formation of solid aerosols by gas phase reactions (Amelin, 1967).

Homogeneous nucleation can be thought to take place in three steps. First, the vapor must be supersaturated to an extent that condensation will take place; second, small clusters of molecules or *embryos* must form; and finally, the vapor must condense on these embryos so that the embryo grows into a full-fledged nucleus which subsequently becomes a droplet. For heterogeneous nucleation only two steps take place, the first and last.

SATURATION RATIO

The saturation ratio of a vapor in a gas can be given by the equality

$$S \equiv \frac{P}{P_\infty(T)} \qquad (12.1)$$

where p is the partial pressure of the vapor in the gas and $p_\infty(T)$ is the saturated vapor pressure of the vapor over a plane of the liquid at a temperature T. When $S > 1$ the gas is said to be supersaturated with vapor, when $S = 1$ the gas is saturated, and when $S < 1$ the gas is unsaturated with vapor. For adiabatic expansion of a gas-vapor system, using the first law of thermodynamics the saturation ratio of a gas saturated prior to expansion can be given by the expression (Amelin, 1967)

$$S = \left(\frac{V_2}{V_1}\right)^{-K} \exp\left(\frac{B}{T_1}\left[\left(\frac{V_2}{V_1}\right)^{K-1} - 1\right]\right) \qquad (12.2)$$

where V_1 and V_2 are the volumes before and after expansion, T_1 is the gas temperature in degrees Kelvin prior to expansion, K is the ratio of the constant pressure specific heat to the constant volume specific heat, and B is a coefficient which comes from the integrated term of the Clausius-Clapeyron equation.

Over a temperature range of $-20°C$ to $60°C$ for water vapor, K has a value of 1.4 and B a value of 5367. Table 12.1 lists K values for several other vapors in addition to water. The term B comes from the equation for approximating vapor pressure:

$$\ln P_\infty(T) = A - \frac{B}{T}$$

Values for A and B are given in Table 13.2.

Table 12.1   Constants for Eq. (12.2) for selected vapors

Material	K	B	Temp Range, °C
Water	1.4	5367	−20 to +60
Ethyl alcohol	1.37	5200	−20 to +100

EXAMPLE 12.1

In an experiment, a chamber holding air saturated with water vapor is rapidly expanding adiabatically to 1.25 times its volume. Determine the value of S following the expansion. The initial temperature, T_1, is $0°C$.

$$S = \left(\frac{V_2}{V_1}\right)^{-K} \exp\left(\frac{B}{T}\left[\left(\frac{V_2}{V_1}\right)^{K-1} - 1\right]\right)$$

$$S = \left(\frac{1.25V_1}{V_1}\right)^{-1.4} \exp\left(\frac{5367}{273}\left[\left(\frac{1.25V_1}{V_1}\right)^{0.4} - 1\right]\right)$$

$$S = (0.732)\exp([19.52][0.093])$$

$$S = (0.732)(6.22)$$

$$S = 4.56$$

It has long been recognized that a small droplet will evaporate even when the gas surrounding it is fully saturated. Supersaturation of the gas is necessary to maintain the drop in equilibrium. Supersaturation is required because the probability of a net loss of a molecule from a convex surface is greater than the probability of net loss from a flat surface of infinite extent. A molecule that has left a small spherical droplet has a much more difficult time finding its way back than it would in finding its way back to a flat surface of infinite extent. Thus the high supersaturations necessary for spontaneous condensation are related to the size of the drop produced.

From cloud chamber studies it was found that with dust-free air, expansion ratios of about 1.37 or so were required for cloud formation to take place. Expansion ratios in this range imply saturation ratios or supersaturations in the order of 700 to 800%. There have been a number of theories advanced to explain the process of self-nucleation, and although none is completely acceptable in all cases, theory is sufficiently adequate to permit a prediction of aerosol parameters for practical applications.

HOMOGENEOUS NUCLEATION – KELVIN'S EQUATION

Consider the energy balance of a nucleating (or condensing) drop. As the droplet (or embryo) is formed, its surface free energy goes from 0 to $\pi d^2 \gamma$ where d is the diameter of the drop and γ is the liquid surface tension. If the free energy potential per molecule is ϕ_a in the vapor phase and ϕ_b in the liquid phase, and n is the total number of molecules contained in the drop growing to a diameter d, then the total change in free energy, ΔG, of the droplet is

$$\Delta G = (\phi_b - \phi_a)n + \pi d_2 \gamma \tag{12.3}$$

Now suppose the partial pressure of the vapor near the droplet is changed by a small amount, dp (keeping the temperature constant). This produces a corresponding change in the free energy per molecule of vapor, $d\phi_a$, and in the free energy per molecule of droplet, $d\phi_b$. If V_a is the volume occupied per molecule in the vapor phase and V_b the volume occupied per molecule in the liquid phase,

$$d\phi_a = V_a dp$$

and

$$d\phi_b = V_b dp$$

Since $V_a \gg V_b$,

$$d\phi_b - d\phi_a = V_a dp = d(\phi_b - \phi_a) \approx \frac{kT}{p} dp \qquad (12.4)$$

In this expression k is Boltzmann's constant.

Integrating Eq. (12.4) with the pressure varying from $p_\infty(T)$ to p gives

$$\phi_b - \phi_a = kT \ln[p/p_\infty(T)] = kT \ln S \qquad (12.5)$$

The mass of a spherical drop is $(\pi/6)d^3\rho$. Hence the number of molecules, n, in the drop is

$$n = \frac{N_A}{M} \frac{\pi}{6} d^3 \rho \qquad (12.6)$$

where N_A is Avagadro's number and M is the molecular weight of the liquid making up the drop. Substituting Eqs. (12.5) and (12.6) in Eq. (12.3) gives

$$\Delta G = \pi d^2 \gamma - [kT \ln S] \left[\frac{N_A}{M} \left(\frac{\pi}{6} d^3 \rho \right) \right] \qquad (12.7)$$

an expression for the total free energy change of the droplet both as a function of drop size and as a function of the saturation ratio.

EXAMPLE 12.2

Compute the free energy elevation for a water droplet of diameter 10Å when $S = 4$. Assume $\gamma = 72$ dynes/cm, $\rho = 1$ g/cm^3, $T = 0°C$.

$$\Delta G = \pi d^2 \gamma - [kT \ln S] \frac{N_A}{M} \frac{\pi}{6} d^3 \rho$$

$$= (3.14)(10^{-7})^2(72) - [(1.38 \times 10^{-16})(273) \ln 4]$$

$$\times \frac{6.02 \times 10^{23}}{18} \frac{(3.14)}{6} (10^{-7})^3 (1)$$

$$\Delta G = 2.26 \times 10^{-12} - 9.15 \times 10^{-13} = 1.35 \times 10^{-12} \text{ ergs}$$

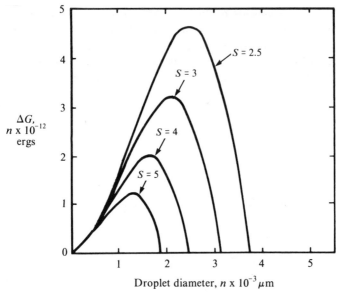

Figure 12.1. Plot of free energy change as a function of particle diameter for various saturation ratios.

Figure 12.1 shows a plot of ΔG as a function of the particle diameter for various values of S. It can be seen from this plot that Eq. (12.7) implies the existence of an energy barrier that acts to prevent the growth of droplets smaller than some critical size. Drops greater than this critical size will continue to grow, since with each slight increase in size the free energy of the system decreases (i.e., the droplet gives up energy). On the other hand, drops smaller than the critical size evaporate, since with these very small drops evaporation reduces their free energy.

The critical drop size can be determined by differentiating G with respect to d, setting the result equal to zero and solving for d. This gives

$$d^* = \frac{4\gamma M}{\rho RT \ln S} \tag{12.8}$$

where R is the universal gas constant. When ρ is in g/cm^3, γ in ergs/cm^2 (or dynes/cm) and the temperature is in $^\circ K$, R has a value of 8.3144×10^7 ergs/$^\circ K$ per gram mole.

Rearranging terms gives Kelvin's equation

$$\ln S = \frac{4\gamma M}{\rho RT d^*} \tag{12.9}$$

A plot of Kelvin's equation is given in Fig. 12.2

EXAMPLE 12.3

Compute the value of S for a droplet diameter, d^*, of 0.01 μm. Assume water at 0°C (γ = 72 dynes/cm, M = 18, ρ = 1 g/cm^3).

$$\ln S = \frac{4\gamma M}{\rho R T d^*}$$

$$\ln S = \frac{4(72)(18)}{(1)(8.31 \times 10^7)(273)(10^{-6})}$$

$$\ln S = 0.229$$

$$S = 1.26$$

The curve shown in Fig. 12.2 is an equilibrium line. If for a given drop of diameter d the value of S associated with it produces a point lying to the left of the line, the drop will evaporate. If the point lies to the right of the line the drop will grow. It is not necessary for a drop of a given size to be associated with a value of S which places a point directly on the curve. The curve indicates the conditions of S and d under which a droplet will evaporate or grow. According to Kelvin's equation, a pure liquid drop will always evaporate when $S < 1$, that is, for water drops in air, if the relative humidity of the air is less than 100%. Even with supersaturation, droplets smaller than the critical size will also evapo-

Figure 12.2. Plot of Kelvin's equation, Equation (12.9).

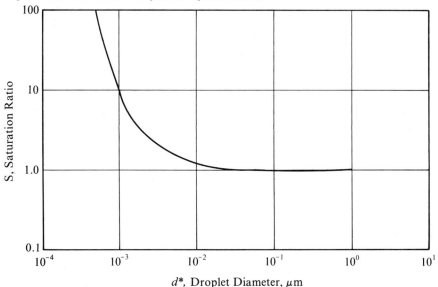

d^*, Droplet Diameter, μm

rate. This implies that small droplets of pure liquids have short lifetimes under normal circumstances. With a monodisperse cloud containing many small drops lifetimes would be longer since the evaporation of some drops results in increased supersaturation leading to growth of other drops.

In determining droplet free energy, it was assumed that the bulk value for surface tension was applicable to all droplet diameters. When the drop is very small it is difficult to envision the meaning of surface tension as it is usually defined, and this point is still the subject of much scientific speculation (Sutugin, 1969). Some authors still consider the use of bulk values for very small droplets to be appropriate (Mason, 1971).

RATE OF FORMATION OF CRITICAL NUCLEI

As mentioned earlier, experiments indicate that spontaneous condensation is not significant until fairly high supersaturations are achieved. For example, supersaturations of about 6 are necessary with water vapor in particle-free air for the formation of a visible fog by adiabatic expansion of moist air at room temperature. This supersaturation implies critical droplet diameters of about 0.001 μm, a cluster of about 50 molecules.

Nucleation embryos for homogeneous nucleation are aggregates of vapor molecules which are constantly being formed and disintegrated by random processes. When a cluster is formed that exceeds the critical size it grows; the likelihood of its formation is a function of the degree of supersaturation. A number of expressions have been derived for the number of clusters reaching critical size per unit time and these have been listed by Pruppacher and Klett (1978). Table 12.2 shows an estimate of the number of drops produced per cubic centimeter per second for various saturation ratios.

From Table 12.2 it appears that a value of $J \approx 1$ is necessary for spontaneous condensation to occur.

Table 12.2   Nucleation rates and molecules per drop at several saturation ratios[a] (Pruppacher and Klett [1978], p. 175)

S	2	3	4	5	6
J (drops/ cm^3-sec)	1.9×10^{-12}	7×10^{-31}	1.1×10^{-10}	7.1×10^{-2}	6.0×10^3
Molecules per drop to achieve growth	889	226	122	72	52
Diameter of drop, μm	3.8×10^{-3}	2.4×10^{-3}	1.8×10^{-3}	1.6×10^{-3}	1.4×10^{-3}

[a]H. R. Pruppacher and J. D. Klett, *Microphysics of Clouds and Precipitation,* D. Reidel Pub. Co., Boston, 1978, p. 175.

In aerosol development, formed nuclei and embryos compete for available vapor molecules. The depletion of vapor caused by the growth of small droplets reduces supersaturation halting nucleation. There have been a number of attempts to model aerosol formation during the expansion of a gas containing a condensable substance (Amelin, 1967), but most require simplifying assumptions and predict droplet number and mean size, saying nothing about the resulting aerosol size distribution. The effect of droplet coagulation is usually neglected, though it is coagulation that leads to the variety of particle sizes formed, and not condensation alone (Fox, Kuhlman, and Reist, 1976).

IONS AS NUCLEI

As mentioned earlier, C. T. R. Wilson (1897) observed that condensation of water droplets in dust-free air took place at lower expansion ratios when ions were present. As charge is placed on a droplet the free energy of that surface is increased approximately by a factor

$$\frac{q_2}{d}\left(\frac{1}{\epsilon_0} - \frac{1}{\epsilon}\right)$$

where q is the charge on the droplet or ionic cluster, ϵ_0 and ϵ are the dielectric constants of the gaseous medium and liquid respectively, and d is the droplet diameter. The total change in free energy becomes

$$\Delta G = -\frac{\pi}{6}d^3\rho\frac{RT}{M}\ln S + \pi d^2\gamma + \frac{q^2}{d}\left(\frac{1}{\epsilon_0} - \frac{1}{\epsilon}\right) \tag{12.10}$$

Similar to the case for homogeneous nucleation, Eq. (12.10) can be differentiated, set equal to zero, and used to determine an expression for the saturation ratio at critical drop diameter:

$$\ln S = \frac{M}{RT\rho}\left[\frac{4\gamma}{d} - \left(\frac{2q^2}{\pi d^4}\right)\left(\frac{1}{\epsilon_0} - \frac{1}{\epsilon}\right)\right] \tag{12.11}$$

Equation (12.11) is plotted in Fig. 12.3 for a single charge. Unlike the pure solution case, droplets carrying charges can exist even at saturation ratios less than one (relative humidities less than 100%). Under these circumstances the droplet size is quite small.

EXAMPLE 12.4

Determine the equilibrium droplet diameter for a water droplet containing a single charge at 80% relative humidity. Assume $T = 70°F$.

$$\ln S = \frac{M}{RT\rho}\left[\frac{4\gamma}{d} - \left(\frac{2q^2}{\pi d^4}\right)\left(\frac{1}{\epsilon_0} - \frac{1}{\epsilon}\right)\right]$$

if $\epsilon_0(\text{air}) = 1.00$, and $\epsilon(\text{water}) = 80.00$

$$\ln(0.8) = \frac{(18)}{(8.31 \times 10^7)(294)(1)} \left[\frac{4(72)}{d} - \left(\frac{2(4.8 \times 10^{-10})^2}{\pi d^4}\right)\left(\frac{1}{1} - \frac{1}{80}\right)\right]$$

$$-3.03 \times 10^8 = \frac{288}{d} - \frac{1.45 \times 10^{-19}}{d^4}$$

$$d = 7.75 \times 10^{-8} \text{ cm} = 7.75\text{Å}$$

Similar to Fig. 12.2, Fig. 12.3 represents a plot of equilibrium values; droplets can be changing in size either away from or toward the equilibrium line. Three distinct cases are possible for condensation or evaporation on an ion, represented by lines A, B, or C on the curve in Fig. 12.3. To determine whether a droplet will grow or evaporate at a given S it is helpful to refer to Fig. 12.4, a plot of the equation for the free energy change on a nucleating ion, Eq. (12.10).

Figure 12.4a shows a free energy plot for case A. Since $\ln S$ is always negative for $\ln S < 1$, the first term in Eq. (12.10) will always be positive. However, when d is very small the $1/d$ term dominates. As d increases the importance of the $1/d$ term decreases while the importance of the d^3 term becomes more

Figure 12.3. Saturation ratio for water as a function of critical particle diameter, single ion, atmospheric pressure: $T = 273°C$.

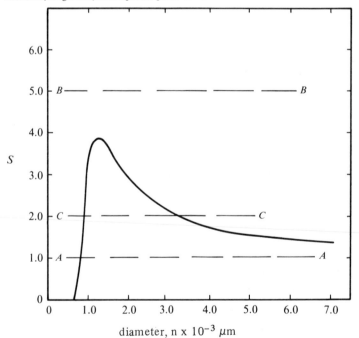

diameter, $n \times 10^{-3} \ \mu m$

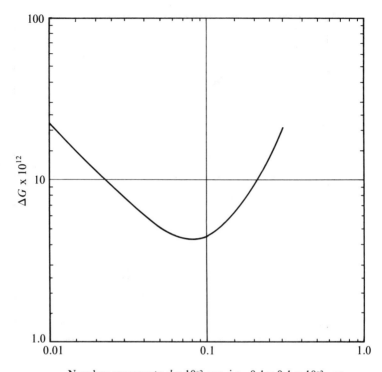

Number represents d x 10^{-2} μm, i.e., $0.1 = 0.1$ x 10^{-2} μm

Figure 12.4a.   Free energy change as a function of drop diameter for droplet containing single ion (line A of Figure 12.3).

apparent (the d^2 term also increases but not to as great a degree). Finally a minimum is reached and subsequently ΔG increases for all increasing values of d. Therefore, droplets whose S value places them along line A will either grow or evaporate toward the equilibrium line, since in this case the equilibrium line represents a stable position.

Case B describes the condition where S is such that the curve of Fig. 12.3 is not intersected at all. The third term in Eq. (12.13) still dominates when d is small, but since the first term is always negative, ΔG is always decreasing (Fig. 12.4b). This means that when S exceeds the maximum value as given in Fig. 12.3, any size charged droplet will grow. This explains the formation of clouds in a cloud chamber. The diameter at which this maximum in Fig. 12.3 occurs is given by:

$$d = \left[\frac{2q^2 \left(\dfrac{1}{\epsilon_0} - \dfrac{1}{\epsilon} \right)}{\gamma \pi} \right]^{1/3} \tag{12.12}$$

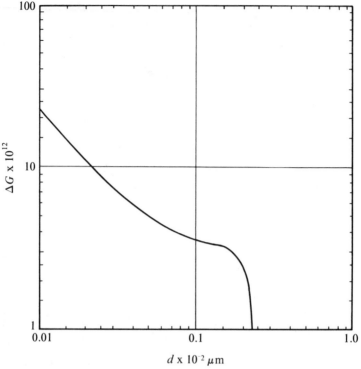

Figure 12.4b.   Free energy change as a function of drop diameter for droplet containing single ion (line *B* of Figure 12.3).

EXAMPLE 12.5

Compute the particle diameter at which S in Fig. 12.3 is maximum. Assume a water droplet at $0°C$, 760 mm Hg pressure.

From Eq. (12.12)

$$d = \left[\frac{2q^2 \left(\frac{1}{\epsilon_0} - \frac{1}{\epsilon} \right)}{\gamma \pi} \right]^{1/3}$$

$$= \left[\frac{2(4.8 \times 10^{-10})^2 \left(\frac{1}{1} - \frac{1}{80} \right)}{(72)(3.14)} \right]^{1/3}$$

$$d = 1.26 \times 10^{-7} \text{ cm} = 12.6Å$$

Finally, there is case *C*, representing a combination of cases *A* and *B*. The free energy shows a minimum and then a maximum (Fig. 12.4c) as particle diameter

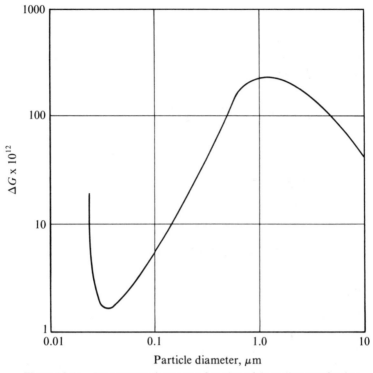

Figure 12.4c. Free energy change as a function of drop diameter for droplet containing a single ion (line C of Figure 12.3).

increases. The minimum could be considered a metastable point. Drops tend to grow or evaporate toward this point and away from the maximum free energy. Hence drops at saturation ratios that place them above the curve in Fig. 12.3 will always grow while those lying below the curve will always evaporate.

There has been good gross experimental verification of Eq. (12.11). For example, droplets formed under conditions where the maximum saturation ratio is just exceeded will continue to grow without bound and become easily visible, while those formed when S is less than this maximum will not be seen. Experimentally, the peak saturation ratio in Fig. 12.3 has been found to be about 4.2 for water condensing on negative ions, in good agreement with theory, but much greater, about 6, for water condensing on positive ions. One possible explanation for the difference in the behavior of positive and negative ions has to do with the orientation of the water molecule. Since the water molecule dipole is thought to have its negative end oriented outward in the outer several layers of the droplet surface, a negative nucleus would permit the capture of water molecules in the correct orientation, whereas with a positive nucleus the molecules would have to turn themselves around before capture, so that condensation, in this case, would be more difficult.

EXAMPLE 12.6

Determine the maximum saturation ratio that corresponds to the diameter in Ex. 12.5. Hence, predict the saturation ratio at which spontaneous condensation on ions will occur. Assume $T = 0°C$. From Eq. (12.11)

$$\ln S = \frac{M}{RT\rho}\left[\frac{4\gamma}{d} - \left(\frac{2q^2}{\pi_d 4}\right)\left(\frac{1}{\epsilon_0} - \frac{1}{\epsilon}\right)\right]$$

$$= \frac{(18)}{8.31 \times 10^7(273)(1)}\left[\frac{4(72)}{1.26 \times 10^{-7}} - \frac{2(4.8 \times 10^{-10})^2}{(3.14)(1.26 \times 10^{-7})^4}\left(1 - \frac{1}{80}\right)\right]$$

$$\ln S = (7.93 \times 10^{-10})(1.71 \times 10^9) = 1.36$$

$$S = 3.89$$

HETEROGENEOUS NUCLEATION

Condensation Nuclei

In most practical cases condensation of a vapor takes place in the presence of small dust particles making unnecessary the extremely high supersaturations required for homogeneous condensation. These small dust particles are given the generic name of condensation nuclei, and can range in size from near molecular sizes to particles greater than 1 μm. A descriptive classification of these nuclei as often used in atmospheric physics is shown in Table 12.3. Although this classification is arbitrary, it corresponds roughly to particle size ranges for different measurement techniques usually employed.

In the atmosphere, small condensation nuclei greatly exceed the number of large ones, with particle number decreasing roughly as the inverse of the cube of particle diameter. Number concentration and particle size are influenced by such factors as topography, meteorology, elevation, vegetative cover, density of human habitation and degree of industrialization. In addition, there are also diurnal as well as seasonal variations in condensation nuclei levels.Concentrations are also influenced by wind speed and direction. Typical outside air condensation

Table 12.3. Condensation nuclei size classifications commonly used in atmospheric physics[a]

Name	Diameter range, μm
Aitken nuclei	0.001–0.4
Large nuclei	0.4 –2.0
Giant nuclei	2.0

[a]C. Junge, *Tellus,* **5**, 1 (1953).

nuclei concentrations can range from as low as 100 particles/cm^3 to 10^6 particles/cm^3 or even higher.

Sources of Condensation Nuclei

Condensation nuclei come from a variety of sources. Such processes as photo-oxidation of natural organic materials over non-urban areas has been suggested as a possible reason for the occurrence of the blue haze usually observed over vegetated areas, (Went, 1960), and although a great deal of work remains to be done to explain the mechanisms of photochemical production of aerosols, it is clear that these reactions are also of great importance in the production of aerosols over urban areas (Goetz and Pueschel, 1967). Photooxidation of organic material may be the most important natural source of condensation nuclei. Other important sources include entrainment of dust particles by the wind, the production of sodium chloride nuclei from sea salt spray, or such spectacular occurrences as forest fires, man-made explosions, or volcanic eruptions. For example, the eruption of Krakatoa in 1883 released a reported 6.5 cubic kilometers of fine dust (Cadle, 1966), equivalent to approximately 10^{23} particles of 0.1-μm diameter. Meteors and interplanetary dust have also been listed by some authors as sources of condensation nuclei. And finally, organic material, both living and dead (plant spores, microorganisms, feathers, skin tissue, hair, etc.) can also act as condensation sites.

EXAMPLE 12.7

The eruption of Mount St. Helens in May of 1980 resulted in the aerosolization of one cubic mile of mountain top. If the average density of the material aerosolized was 2.6 g/cm^3, and 0.1-μm diameter spheres were produced, determine the number of particles produced.

Volume of material aerosolized:

$$V_A = 1 \text{ mi}^3 = (5280 \text{ ft/mi} \times 30.5 \text{ cm/ft})^3$$
$$= 4.18 \times 10^{15} \text{ cm}^3.$$

Volume of one 0.1-μm diameter sphere:

$$V_r = \frac{\pi}{6} d^3 = \frac{\pi}{6} (10^{-5} \text{ cm})^3 = 5.24 \times 10^{-16} \text{ cm}^3$$

Number of particles produced:

$$\frac{V_A}{V_p} = 7.98 \times 10^{30} \text{ particles}$$

If only 1% of the particles were 0.1 μm in diameter, there would have been 7.98 $\times$ 10^{28} particles of this size produced. This is still a lot of particles.

Composition of Condensation Nuclei

Condensation nuclei can be of organic or inorganic composition, can be soluble or insoluble, or can be insoluble with a thin soluble coating (in which case they are termed mixed nuclei). Because of the variety of soluble material existing in the atmosphere, the chemical composition of nuclei is not well defined. Studies of Los Angeles smog collected by electrostatic precipitation indicated that about 60% was made up of inorganic substances or minerals, and the remaining 40% was a complex mixture of organic compounds, carbon, and pollen (Billings et al., 1980). These percentages would not be the same everywhere. However, a great difficulty in analyzing composition is the relatively small mass of material available for analysis — mass contents in a specific size range of 10 µg or less per cubic meter of air are usual. And there may be different chemical fractions for various size ranges of particles. For example, Junge (1963) found that most of the nuclei with diameters between 0.4 and 2 µm collected in Germany and on the East Coast of the United States consisted mainly of ammonium sulfate, whereas the particles whose diameters exceeded 2 µm had a less specific chemical composition, sometimes containing considerable amounts of sodium chloride or sodium nitrate.

Adsorption of atmospheric gases on condensation nuclei can also alter their chemical composition. The pickup of radioactive gases by small particulates is only one form of adsorption, but one which can be easily observed. The exact role of aerosols in the adsorption of gases is an area where little is known at the present time.

Utilization of Nuclei

It should be kept in mind that not all of the atmospheric aerosol enters into the condensation process. In fact, it is only a small fraction of the total. As might be expected from reference to Fig. 12.2, the largest (and most soluble) nuclei are activated first, followed by nuclei of smaller sizes. Thus the utilization of nuclei for condensation depends to a large extent on the degree of supersaturation present, and in the atmosphere this, in turn, depends on the rate of cooling of the air.

Utilization also depends on the chemical composition of the nuclei. There are two general classes of condensation nuclei to be considered — soluble nuclei and insoluble nuclei. With soluble nuclei the condensing vapor dissolves the nucleus, changing the properties of the embryo drop from that of a pure liquid. With insoluble nuclei surface characteristics are important, since once the nucleus is coated with liquid it behaves in a manner similar to a pure liquid drop.

Insoluble Nuclei

The two extremes of insoluble nuclei are nuclei which are easily wetted and those which are not. Nuclei which are easily wetted rapidly take on the appearance of a droplet and subsequently behave like one. To predict droplet growth or evaporation these particles with easily wettable surfaces can be considered to be pure drop nuclei, and the Kelvin equation can be used directly (but with a lower limit on nucleus size).

In cases where the particle surfaces are not wettable, condensation proceeds with much more difficulty. This is because the condensing liquid tends to pull into small spheres on the particle surface and only when the entire surface is covered with these spheres is a liquid coating formed. Fletcher (1958a, 1958b) has treated this problem by considering the contact angle between an embryo sphere formed on the particle and the particle surface. His results correspond to what has been observed experimentally — that it is very difficult to get condensation to take place on non-wettable particles unless high supersaturations are used. The role of insoluble nuclei in the condensation process is still in question and remains another problem for future investigators to resolve.

Soluble Nuclei

In many instances condensation takes place on soluble nuclei, producing solution droplets. An example is the condensation of water on a sodium chloride nucleus. Initially a saturated NaCl solution is formed. As condensation proceeds further, the solution becomes more and more dilute until finally the drop behaves in a manner similar to a droplet of pure liquid. In general the equilibrium solvent vapor pressure over a solution surface is lower than over a pure solvent surface, the amount of decrease depending on the nature of the solvent and the concentration and nature of the solute. A lower equilibrium vapor pressure means that condensation occurs at lower saturation ratios.

For an electrolyte solution Robinson and Stokes (1959) give for the reduction in equilibrium vapor pressure the relationship

$$\frac{p_\infty'(T)}{p_\infty(T)} = \exp\left(- v \S \vartheta M\right) \tag{12.13}$$

where $p_\infty'(T)$ is the equilibrium vapor pressure over an infinite plane of solution, $p_\infty(T)$ is the equilibrium vapor pressure over an infinite plane of pure solvent, M is the molecular weight of the solvent, $\S$ is a coefficient, known as the molal osmotic coefficient, and v is the number of ions per molecule available for complete ionization. Table 12.4 gives values of $\S$ and v for several electrolytes at various concentrations. The factor ϑ is the number of moles of solute per gram of solvent, which for a single spherical drop can be given by

$$\vartheta = \frac{m}{W}\left(\frac{\pi}{6} d^3 \rho' - m\right)^{-1} \tag{12.14}$$

where W is the molecular weight of the solute, m the mass of the solute per drop and d is the drop diameter. The primed values refer to the solute-solvent mixture and unprimed values to the pure materials.

Table 12.4. Osmotic coefficients, §, of some electrolytes at 25°C[a]

Molality	NaCl $v = 2$	MgCl $v = 3$	$(NH_4)_2SO_4$ $v = 3$	$Ca(NO_3)_2$ $v = 3$	$Al_2(SO_4)_3$ $v = 5$
0.1	0.932	0.861	0.767	0.827	0.420
0.2	0.925	0.877	0.731	0.819	0.390
0.4	0.920	0.919	0.690	0.821	0.421
0.6	0.923	0.976	0.667	0.831	0.545
0.8	0.929	1.036	0.652	0.843	0.718
1.0	0.936	1.108	0.640	0.859	0.922
1.2	0.943	1.184	0.632	0.879	
1.6	0.96	1.347	0.624	0.917	
2.0	0.983	1.523	0.623	0.917	
2.5	1.013	1.762	0.626	1.001	
3.0	1.045	2.010	0.635	1.051	
3.5	1.080	2.264	0.647	1.103	
4.0	1.116	2.521	0.660	1.157	
5.0	1.192	3.048	0.686	1.263	
5.5	1.231		0.699	1.313	
6.0	1.272			1.361	

[a]Abridged from R. A. Robinson and R. H. Stokes, *Electrolyte Solutions*, Butterworth, London, 1959, p. 483.

Recalling Eq. (12.9), for a pure droplet

$$\ln[S] = \frac{4\gamma M}{\rho R T d^*}$$

This can be rewritten as

$$\frac{p}{p_\infty(T)} = \exp\left(\frac{4\gamma M}{\rho R T d^*}\right) \qquad (12.15)$$

For a solution droplet Eq. (12.15) can be written

$$\frac{p'}{p_\infty'(T)} = \exp\left(\frac{4\gamma' M}{\rho' R T d^*}\right) \qquad (12.16)$$

Combining Eq. (12.16) and (12.13) gives

$$\frac{p'}{p_\infty(T)} = p'/p_\infty'(T)\frac{p_\infty'(T)}{p_\infty(T)} = \exp\left(\frac{4\gamma' M}{\rho' R T d^*} - v\phi m M\right) \qquad (12.17)$$

an expression for the ratio of the vapor pressure above a solution droplet (see Byers, 1965a or Byers 1965b).

Close examination of various equations that have been proposed for predicting saturation ratios over solution droplets reveals that they differ only in detail and all give essentially the same results. Figure 12.5 is a plot of S versus d^* for NaCl masses of various sizes with water as the solvent, using Eq. (12.17). Curves similar to these are very often referred to as "Koehler curves."

EXAMPLE 12.8

Determine the value of $p'/p_\infty(T)$ for a 0.01-μm $(NH_4)_2SO_4$ particle when $d*$ is 0.1 μm ($T = 20°C$). Assume a spherical shape, $\rho = 1.77$ g/cm³, $W = 132$, $M = 18$.

$$\frac{p'}{p_\infty(T)} = \exp\left(\frac{4\gamma'M}{\rho'RTd*} - v\S\vartheta M\right)$$

From Eq. (12.14)

$$\vartheta = \frac{m}{W} \Big/ \left(\frac{\pi}{6}d^3\rho' - m\right)$$

$$\rho' = \frac{\text{wt. solute} + \text{wt. solvent}}{\text{vol. solute} + \text{vol. solvent}}$$

$$= \frac{\frac{\pi}{6}(0.01 \times 10^{-4})^3 \times 1.77 + \left(\frac{\pi}{6}d*^3(1) - \frac{\pi}{6}(0.01 \times 10^{-4})^3(1)\right)}{\frac{\pi}{6}d*^3}$$

$$\rho' = \frac{(0.01 \times 10^{-4})^3(0.77)}{10^{-15}} + 1 \cong 1.00$$

Assume $' = \gamma = 72$

$$m' = \frac{\pi}{6}(0.1 \times 10^{-4})^3(1) = 5.24 \times 10^{-16} \text{ g}$$

$$m = \frac{\pi}{6}(0.01 \times 10^{-4})^3(1.77) = 9.27 \times 10^{-19} \text{ g}$$

$$\text{Molality} = \frac{(9.27 \times 10^{-19}/132) \times 1000}{5.24 \times 10^{-16}} = 1.34 \times 10^{-2}$$

From Table 12.4: $\S = 0.7, v = 3$

$$\frac{p'}{p_\infty(T)} = \exp\left[\frac{4(72)(18)}{(1)(8.31 \times 10^7)(293)(10^{-5})} - (0.7)(3)(1.34 \times 10^{-5})(18)\right]$$

$$= \exp[2.13 \times 10^{-2} - 5.06 \times 10^{-4}]$$

$$\frac{p'}{p_\infty(T)} = 1.02$$

Note that this value is the same as for the pure droplet case. There is a difference only when the drop size is close to the nucleus size.

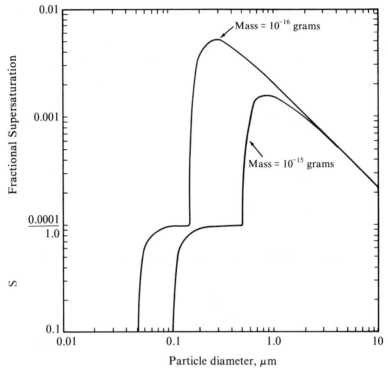

Figure 12.5. Plot of saturation ratio or supersaturation as a function of critical particle diameter for soluble nuclei of 10^{-15}g and 10^{-16}g.

Unlike the curve for condensation on a droplet of pure solvent, when a solute is present it is possible to have condensation taking place even at relative humidities of less than 100% [when $p'/p_\infty(T) < 1$]. The effect of a solute can be considered to be very similar to the effect of an ion on droplet growth or evaporation except that the basic nucleus size can be much larger.

Analogous to the case for condensation on ions, at a given $p'/p_\infty(T)$ droplets will grow or evaporate away from the portion of the curve to the right of the maximum and toward that portion lying to the left of the maximum unless $p'/p_\infty(T)$ is so great that they grow without bound. As a result it is possible to have stable solution droplets whose sizes are only a function of the mass of solute and the ratio $p'/p_\infty(T)$. For example, a 1-μm-diameter droplet containing 10^{-15} g of NaCl will rapidly evaporate to a diameter of about 0.6 μm in an atmosphere where $p'/p_\infty(T)$ is 1.001, whereas if it were initially 3 μm in diameter on formation, the drop would grow without bound until it eventually depleted the water vapor around it or was removed by some process such as sedimentation.

The injection of soluble particles into humid air results in the almost immediate generation of stable droplets of a much larger size. Figure 12.6 shows a plot

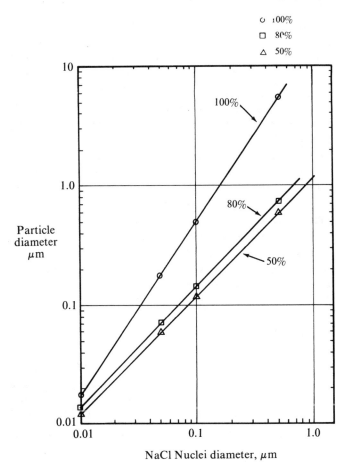

Figure 12.6. Stable droplet diameter as a function of soluble nuclei diameter (NaCl) for various relative humidities.

of stable droplet diameter as a function of NaCl particle diameter (assuming spherical particles) for various relative humidities. At 100% relative humidity particle size is increased about five times for NaCl masses of about 10^{-15} g and about 10-fold for masses of about 10^{-13} g. It is this increase in particle size that is responsible for the evolution of haze in the atmosphere when adequate numbers of soluble nuclei are present in conjunction with high humidities. Unlike completely pure droplets, because of hysteresis effects, slight changes in humidity will not significantly alter stable drop size for some solutes.

Hysteresis in Evaporation and Condensation

Hysteresis describes a process in which a phase change occurs at one humidity when the humidity is rising with the reverse change not occurring at the same humidity value but at some different humidity when the humidity is falling. A soluble hygroscopic particle in an atmosphere of vapor-laden solvent will initially pick up a solvent envelope by adsorption. At some minimum "relative humidity" the quantity adsorbed becomes such that the soluble particle is dissolved and becomes a liquid droplet. If the humidity is reduced to dry the droplet, it has been observed that the drop remains a liquid even at relative humidities less than that required for initial solution, implying supersaturation of the solution making up the drop. With continued reduction of the "relative humidity" the solute in the drop suddenly crystallizes (Fig. 12.7).

This hysteresis effect was studied by Orr and his colleagues (1958), who found that solution takes place over a range of 68% to 80% relative humidity for various inorganic salts, while recrystallization does not occur until relative humidities are about 30% lower. For example, for NaCl solutions, crystallization occurs at a relative humidity of 70% and recrystallization appears at 40%. For nonhygroscopic materials, the effect does not occur. This phenomenon helps explain why smogs and hazes persist at relative humidities well below those at which they originally were formed.

PROBLEMS

1. For a 0.01-μm water droplet, compute the value of ΔG when $S = 4$.

2. A volume of air at 80°F at sea level is expanded by rising to an elevation of 750 ft above sea level. If the expansion is adiabatic and the air is initially saturated with water vapor, what is the resulting value of S?

3. Determine the value of S at which a 0.03-μm-diameter water droplet will just continue to grow.

4. It was found by C. T. R. Wilson (1897) that when air at 20°C, initially saturated with water vapor and free of any condensation nuclei, was expanded with an expansion ratio in excess of 1.37, homogeneous nucleation occurred. What is the value of S which is implied by this expansion ratio?

5. Wilson (1897) found that the condensation of water vapor occurred on a negative ion with an expansion ratio of 1.25, whereas for condensation on a positive ion an expansion ratio of 1.31 was necessary. What is the expansion ratio equivalent to the maximum of the S versus d^* plot? (How well does theory agree with experiment?) What is the value of d^* associated with this expansion ratio?

6. A cloud contains 3×10^4 condensation nuclei/cm^3. The condensation nuclei are 0.01-μm-diameter unit density spheres. How many cubic meters of air must be sampled to get 1 mg. of condensation nuclei?

7. What supersaturation is necessary for a 0.05-μm-diameter ammonium sulfate particle (sphere) to grow without limit? What is this value for a 0.5-μm-diameter ammonium sulfate particle?

8. Determine the equilibrium droplet diameter for a water droplet containing three positive charges at 90% relative humidity. Assume $T = 70°$ F.

9. How many particles would have been produced from the eruption of Mt. St. Helens (see Example 12.7) if the average density of the material was 3 g/cm^3 and the particles were 0.01 μm in diameter.

10. A sodium chloride solution droplet has a diameter of 1 μm. If the solubility of sodium chloride in water is 35.7 $g/100$ cc and sodium chloride content of the drop is 25% of the solubility, determine the diameter of the resulting sodium chloride particle if all water is evaporated. Assume spheres in both cases.

THIRTEEN

EVAPORATION AND GROWTH

Besides knowing under what circumstances a droplet will form, it is important to be able to estimate rates of droplet growth or evaporation. For example, soluble particles inhaled into the respiratory system will rapidly become humidified and may grow to such an extent that their aerodynamic properties in the lung are quite different from when they are measured outside of the lung. Or insecticide droplets sprayed from the air with the hope that they will reach the ground may actually evaporate and never carry out their intended purpose.

MAXWELL'S EQUATION

The equation for the evaporation of a droplet as a function of time was first derived by James Maxwell in 1877. Although his derivation contains a number of simplifying approximations, Maxwell's equation gives reasonable results for fairly large droplets of pure substances. For this equation, it is assumed that at a given temperature the partial pressure at the surface of a drop equals the saturation pressure, i.e., $p_{surface} = p_\infty$. In terms of concentrations of molecules, this means that the vapor concentration at the surface just equals the concentration of saturated vapor. This assumption is valid when the drop size is not too small compared to the mean free path of the vapor molecules. Considering that the molecules can leave the drop surface by diffusion,

$$J = -D\frac{\partial c}{\partial R}$$

where J is the number of grams of solvent vapor passing through unit area in unit time and R is a distance measured from the center of the droplet. The term c represents the concentration of solvent vapor (in g/cm^3). The number of grams of vapor lost or gained per second through a spherical surface of radius R enclosing the droplet, I, is

$$I = 4\pi R^2 D\frac{\partial c}{\partial R} \tag{13.1}$$

Integration of Eq. (13.1) gives

$$c = \frac{I}{4\pi RD} + \text{constant} \tag{13.2}$$

where the constant depends on the conditions of the problem. If c_0 is the concentration of vapor a large distance away from the drop, then

$$c \approx \frac{I}{4\pi RD} + c_0 \tag{13.3}$$

For practical purposes the approximation can be considered to be an equality. By the initial assumption that the concentration of vapor at the droplet surface ($R = d/2$) is equal to the saturation concentration, c_∞, then

$$c_\infty = \frac{I}{2\pi dD} + c_0 \tag{13.4}$$

and

$$I = 2\pi Dd(c_\infty - c_0) \tag{13.5}$$

known as Maxwell's Equation. The evaporation rate, I, is a function of drop diameter, d, diffusion coefficient of the solvent vapor, D, and the difference between the solvent vapor pressure and ambient partial pressure of the vapor. A relationship between vapor concentration and pressure can be found by considering the vapor to be an ideal gas. Then

$$c = \frac{pM}{RT} \tag{13.6}$$

where, if p is the vapor pressure in mm Hg, M the molecular weight in gram-moles and T the temperature in °K, R has a value of 62,360 cm^2 (mm Hg) °K^{-1}

Table 13.1. Values of diffusion coefficients for various gas mixtures, STP[a]

Gases	D, cm^2/sec
H$_2$O-air	0.219
Ethanol-air	0.099
Ethyl-other-air	0.070
Benzene-oxygen	0.080
Mercury-nitrogen	0.119
Iodine-nitrogen	0.070
Iodine-air	0.069

[a]Compiled from W. Jost, *Diffusion in Solids, Liquids and Gases*, Academic Press, New York, 1952, p. 412.

(mole)$^{-1}$. This gives c the units of g/cm^3. Maxwell's equation can now be written in terms of vapor pressure, that is,

$$I = \frac{2\pi DMd}{RT}(p_\infty - p_0) \tag{13.7}$$

The diffusion coefficient for water vapor in air is 0.22 cm^2 sec^{-1} at 0°C and 760 mm mercury pressure (STP). Values for other air-vapor mixtures are given in Table 13.1. For other temperatures the diffusion coefficient can be estimated by the relationship:

$$D \approx D_0 \left(\frac{T}{T_0}\right)^{3/2} \tag{13.8a}$$

whereas for other pressures,

$$D = D_0 \left(\frac{p}{p_0}\right) \tag{13.8b}$$

The vapor pressure of a liquid at various temperatures can be estimated by:

$$\ln p_\infty = A - \frac{B}{T} \tag{13.9}$$

where T is in °K, p in mm Hg and the constants A and B are given in Table 13.2.

EXAMPLE 13.1

A 30-μm-diameter water droplet is evaporating in a chamber. The chamber temperature is 20°C and pressure is 760 mm Hg. The chamber relative humidity is 50%. Find the droplet evaporation rate in grams of water lost per second.

From Eq. (13.7)

$$I = \frac{2\pi D M d}{RT}(p_\infty - p_0)$$

from Table 13.1 and Eq. (13.8a)

$$D = (0.219)\left(\frac{293}{273}\right)^{3/2}$$

$$= 0.244 \text{ cm}^2 \text{ per sec}$$

from Table 13.2 and Eq. (13.9)

$$\ln p_\infty = 21.18 - \frac{5367.2}{293} = 2.86$$

$$p_\infty = 17.49 \text{ mm Hg}, p_0 = \frac{17.49}{2} = 8.75 \text{ mm Hg}$$

$$I = \frac{(2)(3.14)(0.244)(18)(30 \times 10^{-4})}{(62,360)(293)}(17.49 - 8.75)$$

$$I = 3.96 \times 10^{-8} \text{ g per sec}$$

Table 13.2. Values of the constants A and B for various liquids, $T \approx 20°C$ (for Eq. 13.9)[a]

Liquid	A	B
Bromine	18.47	3915
Water	21.18	5367
Iodine	22.96	7155
Uranium hexafloride	22.95	5344
Acetone	18.51	3906
Benzene	20.60	4759
Bromobenzene	19.44	5364
Decane	19.67	5694
Diethylether	17.97	3488
Ethanol	21.24	5200
n-Hexane	18.09	3896
Methyl iodide	17.54	3439
Naphthalene	25.87	8423
n-octane	23.98	6537

[a] Adapted from F. Daniels and R. Alberty, *Physical Chemistry*, 2nd ed., Wiley and Sons, 1963, p. 126.

Maxwell's equation only holds for condensation or evaporation into air and neglects temperature changes in the drop. If the drop were evaporating into a vacuum, the evaporation rate would be a function of the square of the droplet diameter. With high Knudsen numbers, Eq. (13.5) is not valid without a correction for the free molecular nature of the medium.

GROWTH OR LIFETIME OF DROPS–LANGMUIR'S EQUATION

In the previous section the evaporation rate of a droplet of a specific diameter was considered. But as the drop evaporates its diameter decreases. Since a large number of molecules are required to affect the drop size significantly, a quasi-stationary condition can be assumed to estimate crudely the drying time, or "lifetime" of the drop. Then

$$I = \frac{dm}{dt} \tag{13.10}$$

Since

$$dm = \frac{\pi}{2} d^2 \rho dd \tag{13.11}$$

and

$$\frac{dm}{dt} = \frac{2\pi DMd}{RT}(p_\infty - p_0), \tag{13.12}$$

$$\frac{dd}{dt} = \frac{4DM}{d\rho RT}(p_\infty - p_0). \tag{13.13}$$

Integration and rearrangement of terms gives the time for a droplet of diameter d to evaporate to a diameter of d_0 when the partial pressure in the surrounding medium is p_0. This equation, sometimes known as Langmuir's equation, is

$$t = \frac{\rho RT(d^2 - d_0^2)}{8DM(p_\infty - p_0)} = \frac{\rho RT(d^2 - d_0^2)}{8DMp_\infty(1 - S)} \tag{13.14}$$

where S is the saturation ratio.

Experimental varification of Langmuir's equation is sparse but some data were recently compiled by Davies (1978) which show reasonably good agreement with Eq. (13.14) when temperature corrections are taken into account. Equation (13.14) also can be applied to a condensing aerosol where d_0 is the final droplet diameter. For a polydisperse collection of condensation nuclei (particles > 0.1 μm in diameter) it can be seen from Eq. (13.14) that the rate of growth of small droplets will be faster than the rate of growth of larger droplets. Thus, as long as there is an excess of vapor for condensation, there will be a tendency for condensing droplet size distributions to become more monodisperse. When a polydisperse cloud of droplets dries, the degree of polydispersion is increased both through the differences in drying rates for the various sizes of drops and through enhanced coagulation caused by the dispersity of drop sizes.

EXAMPLE 13.2

Estimate the time required for a 20-μm diameter water droplet to evaporate completely when the relative humidity is 50%. Assume the temperature is 20°C.

$$t = \frac{\rho R T (d_0^2 - d^2)}{8 D M p_\infty (1 - S)}$$

$$t = \frac{(1)(62,360)(293)(20 \times 10^{-4})^2}{(8)(0.244)(18)(17.49)(0.5)}$$

$$t = 0.24 \text{ sec}$$

A difficulty with this type of estimate is that the droplet temperature will be somewhat below air temperature, lowering vapor pressure and thus increasing actual drop lifetime.

MODIFICATIONS TO LANGMUIR'S EQUATION

Calculations using Langmuir's equation show that the lifetime of small volatile droplets in air is surprisingly short. On the other hand, for large drops the lifetimes appear to be approximately correct as indicated by experimental measurements. It is possible to extend the useful range of this equation to somewhat smaller size drops. Fuchs (1959) initially pointed out that Langmuir's equation could not be correct for small particles having diameters approaching the mean free path of the gas since the equation predicted rates of molecular escape exceeding the evaporation rate into a vacuum. To correct for this difficulty, Fuchs considered the diffusion process to start a distance of approximately one mean free path from the droplet surface.

Applying Fuchs' modification to Langmuir's equation in the case of evaporation into a space where $p_0 = 0$ gives

$$t = \frac{\rho R T}{D M p_\infty}\left[\frac{d^2}{8} + \frac{dD}{2\alpha v_x} - \frac{\Delta}{2}d + \Delta^2 \ln\left(\frac{d + 2\Delta}{2\Delta}\right)\right] \tag{13.15}$$

In Eq. (13.15) Δ represents the distance an evaporating molecule must travel before it strikes a "gas" molecule, and can be estimated from

$$\Delta = \lambda\left(\frac{m_1 + m_2}{m_1}\right)^{1/2} \tag{13.16}$$

The term α is an accommodation coefficient which has a value for pure water according to Fuchs (1959) of 0.034 (for water with impurities the value can be even smaller), m_1 and m_2 are the masses of single gas and vapor molecules respectively, λ is the mean free path of the gas molecules, and v_x is

$$v_x = \left(\frac{kT}{2\pi m_2}\right)^{1/2} \tag{13.17}$$

Fuchs gives Eq. (13.17) for velocity, indicating that this is one-quarter of the mean absolute velocity of the vapor molecules. Sedunov (1974) follows Fuchs' derivation but gives for the velocity an expression without the $\sqrt{1/\pi}$. The most probable velocity of a Maxwellian distribution is, from Eq. (3.9), 4 times Sedunov's result.

The effect of Fuchs' modification on Eq. (13.14) is to increase the lifetime of very small drops fairly significantly and even to have some effect on drops having diameters of 10 μm or more. If F_c is defined as the ratio of Eq. (13.15) to Eq. (13.14) for similar conditions, Fig. 13.1 shows a plot of this ratio as a func-

Figure 13.1. Plot of Fuchs (1959) correction; ratio of Equation (13.15) to Equation (13.14).

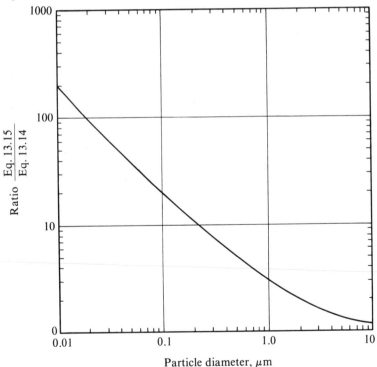

EXAMPLE 13.3

Compute the value of t using Fuchs' correction for a 2-μm water droplet as 20°C (Use $\alpha = 0.034$).

From Eq. (13.15)

$$t = \frac{\rho RT}{DMp_\infty}\left[\frac{d^2}{8} + \frac{dD}{2\alpha v_x} - \frac{\Delta}{2}d + \Delta^2 \ln\left(\frac{d+2\Delta}{2\Delta}\right)\right]$$

$$v_x = \left[\frac{(1.38 \times 10^{-16})(293)(6.05 \times 10^{23})}{(2)(3.14)(18)}\right]^{1/2}$$

$$v_x = 1.47 \times 10^4 \text{ cm/sec}$$

$$\Delta = 0.07 \times 10^{-4}\left(\frac{18+29}{29}\right)^{1/2}$$

$$\Delta = 8.91 \times 10^{-6} \text{ cm}$$

$$t = (2.38 \times 10^5)[5.00 \times 10^{-9} + 4.88 \times 10^{-8} - 8.91 \times 10^{-10} + 1.99 \times 10^{-10}]$$

$$t = (2.38 \times 10^5)(5.31 \times 10^{-8}) = 1.26 \times 10^{-2} \text{ sec}$$

tion of d, the droplet diameter. As an example, for 1-μm-diameter droplets, evaporation times as estimated using the Fuchs modification are about 3.5 times greater than those estimated with Langmuir's equation alone.

A second modification is necessary to account for the change in temperature inside the droplet during condensation or evaporation. This change affects the liquid vapor pressure, decreasing evaporation rates and increasing condensation rates. Mason (1971) gives a modified form of Eq. (13.14) that accounts for temperature effects including elevation or depression of drop vapor pressure as

$$t = \frac{\left[\frac{L\rho}{K_t}\left(\frac{LM}{RT}-1\right) + \frac{\rho RT}{DMp_\infty}\right](d_0^2 - d^2)}{8(1-S)}. \tag{13.18}$$

The term L represents the latent heat of condensation and K_t is the thermal conductivity of the air. These values vary only slightly with temperature and are given in Table 13.3 for several temperatures. The effect of temperature change within the drop during condensation or evaporation depends very strongly on ambient temperature. This is shown in Table 13.4 where the ratio of Eq. (13.18) to Eq. (13.14) is computed for a number of different temperatures. The effect of temperature change is more noticeable at high than at low temperatures.

Table 13.3. Values of the constants K_t and L for water droplets in air at 760 mm

Temperature, °C	K	L
-10	5.64	600
-5	5.72	598
0	5.80	595
5	5.88	593
10	5.96	590
15	6.04	588
20	6.16	585
30	6.28	580
40	6.44	574
50	6.60	568

Table 13.4. Effect of temperature change correction on Langmuir's equation

Temperature difference, °C	Correction factor, F_c
-20	1.15
-10	1.31
0	1.62
10	2.18
20	3.13
30	4.70
40	7.19
50	11.02

The temperature within a drop at steady state (constant evaporation rate and steady temperature) can be given by the expression

$$T_0 - T_\infty = \frac{D_T ML}{RK}\left(\frac{p_\infty}{T_\infty} - \frac{p_0}{T_0}\right) \tag{13.19}$$

The temperature of the drop is independent of drop size. As pointed out by Davies (1978) and others, this equation is the fundamental psychrometric equation that permits wet-bulb temperatures to be calculated. Thus a psychrometric chart can be used to estimate steady-state drop temperature by finding the wet-bulb temperature corresponding to a given ambient temperature and relative humidity. This wet-bulb temperature is also the evaporating droplet temperature.

EXAMPLE 13.4

Estimate the drying time of a 10-μm-diameter water droplet at 40°C taking into account temperature changes within the drop. Let $S = 0$.

Using Eq. (13.18)

$$t = \frac{\left[\frac{L\rho}{K_t}\left(\frac{LM}{RT} - 1\right) + \frac{\rho RT}{DMp_\infty}\right](d_0^2 - d^2)}{8(1 - S)}$$

D and p_∞ are computed from Eqs. (13.8a) and (13.9) respectively for a temperature of 313 °K.

$$D = 0.27 \text{ cm}^2 \text{ per sec}, p_\infty = 56.4 \text{ mm Hg}$$

From Table 13.3 $L = 586$ cal/g, $K_t = 6.09 \times 10^{-5}$ cal cm^{-1} sec^{-1} °K^{-1}.

$$t = [(3.07 \times 10^4)(-1) + 7.12 \times 10^4] \frac{(1.0 \times 10^{-6})}{8}$$

$$t = 5.06 \times 10^{-3} \text{ sec}$$

EVAPORATION TIME IN A SATURATED MEDIUM

Langmuir's equation indicates that a droplet will not evaporate when $S = p/p_\infty = 1.0$. But according to the Kelvin equation, curvature effects will cause small droplets to evaporate, even when S exceeds 1.0. How can this apparent contradiction be resolved? One way is to replace the $(p_\infty - p_0)$ term in Eq. (13.13) with an equivalent term from Kelvin's equation which takes curvature into account. Recalling Kelvin's equation,

$$\frac{p}{p_\infty} = \exp\left(\frac{\gamma M}{\rho R T d}\right) \tag{13.20}$$

which can be written as an expansion

$$\frac{p}{p_\infty} = 1 + \frac{\gamma M}{\rho R T d} + \frac{1}{2}\left(\frac{\gamma M}{\rho R T d}\right)^2 + \frac{1}{6}\left(\frac{\gamma M}{\rho R T d}\right)^3 + \ldots \tag{13.21}$$

If $p/p_\infty \approx 1$, the first two terms of Eq. (13.21) are adequate to determine p/p_∞ so that

$$\frac{p}{p_\infty} = 1 + \frac{\gamma M}{\rho R T d} \tag{13.22}$$

$$p_\infty - p = -\frac{\gamma M p_\infty}{\rho R T d} \tag{13.23}$$

For the case of complete evaporation, $p = 0$, and Eqs. (13.13) and (13.23) can be combined to give

$$\frac{dd}{dt} = \frac{4DM^2\delta p_\infty}{d^2\rho^2R^2T^2} \tag{13.24}$$

which gives, on integration and rearrangement of terms,

$$t = \frac{d^3}{12D\gamma p_\infty}\left(\frac{\rho RT}{M}\right)^2 \tag{13.25}$$

This represents an expression for the total lifetime of droplets in a medium that is essentially saturated with vapor.

EXAMPLE 13.5

Estimate the lifetime of a 0.1-μm diameter water droplet in saturated air at 20°C.

From Eq. (13.25)

$$t = \frac{d^3}{12D\gamma p_\infty}\left(\frac{\rho RT}{M}\right)^2$$

$$t = \frac{(10^{-5})^3}{(12)(0.24)(72)(17.49)(1333)}\left(\frac{(1)(8.314 \times 10^7)(293)}{(18)}\right)^2$$

Note that p had to be converted to units of dynes/cm^2 by the factor 1333 because the surface tension units are also dynes/cm^2 and R is expressed in appropriate units.

$$t = 3.79 \times 10^{-4} \text{ sec}$$

GROWTH AND EVAPORATION OF MOVING DROPLETS

As a droplet moves relative to its surrounding medium, its evaporation rate may be changed by having the medium sweep away vapor molecules near the drop's surface. Fuchs (1959) examined in detail both theory and experimental data on this question and concluded that from a theoretical standpoint in the case of droplets evaporating in the Stokes region (Re>1), increased evaporation on the front face of the drop is balanced by decreased evaporation on the rear. Thus he concluded that the overall rate would be unchanged. In addition, even though at very high Reynolds numbers evaporating molecules could be removed so quickly that evaporation of the drop would be similar to evaporation in a vacuum, for particles smaller than about 40 μm in diameter this never happens, since movement relative to the medium is at a low Re, or the drops are rapidly decelerated to a low Re. Therefore, drop motion can be neglected in considering most evaporation and condensation estimates.

PROBLEMS

1. Determine the lifetime of a 50-μm-diameter benzene droplet when it is released into an atmosphere of oxygen at 25°C.

2. What would be the value of t for this same benzene drop using Fuchs' correction. Make whatever assumptions are necessary.

3. What would be the drying time for a 10-μm water droplet when Mason's equation, Eq. (13.18) is used? Assume an air temperature of 20°C.

4. A 200-μm droplet is evaporating in air at 25°C and 40% relative humidity. At equilibrium, what is the internal temperature of the droplet?

5. Mason (1971) states that droplets with average radii of 30 μm evaporate completely after falling a few meters in unsaturated air. Show that this is true.

6. E. G. Zak (1936) found that the time for complete evaporation of a 1-mm water droplet suspended from a thin glass fiber was 605 sec at 20°C. Compare this value to one computed from
   a. Langmuir's equation
   b. Eq. (13.18)

7. In an experiment water droplets were dried over phosphorus pentoxide to give an atmosphere of 0% relative humidity. Using Eq. (13.19), determine the drop temperature if the ambient temperature is 19°C. Compare this value to one found from use of the psychrometric chart in Appendix F.

FOURTEEN

OPTICAL PROPERTIES
Extinction

INTRODUCTION

Optical effects of aerosols represent the most spectacular and most important of all aerosol characteristics. Clouds, haze, and smokes all appear as they do because of the optical properties of the individual particles and the effects of these particles on each other. Brilliant sunsets inspire, myriad cloud colors excite, and dense, thick fogs disorient, feelings created by the effects of light scattering and absorption.

Most objects that can be seen are visible because they scatter light. A tree is seen with all its shadings of light and dark because of variations in scattering intensities over the different parts of the tree — otherwise only a silhouette or outline of the tree would be seen. Selective scattering and absorption are responsible for colors of things. Thus a leaf on a tree illuminated with natural light looks green because it absorbs red light more efficiently than green light. Black smoke appears black because the smoke particles efficiently absorb all visible wavelengths of light. White smoke appears white because the smoke particles efficiently scatter all visible wavelengths.

All aerosol particles scatter light. Some, principally metallic or carbon particles, or very large particles, can also absorb light. The combination of scattering plus absorption is called extinction.

Scattering functions that describe the light scattered or absorbed by a particle can be computed for spherical or cylindrical shapes using a general mathematical theory known as the *Mie theory* after G. Mie(1908), although other investigators at the time had proposed essentially identical theories [see, for example, the discussion by Kerker,(1969)]. Although Mie's theory was formulated for spherical particles, experiments by Napper and Ottewill(1964) and Berry(1962,1966) indicate that angular scattering patterns and extinction predictions for isometric particles such as cubes or octrahedra differ very little from those for spherical particles of the same equivalent size.

In this chapter overall scattering properties of particles will be examined and the relationship of these properties to visual phenomena reviewed. In Chapter 15 angular scattering of light from aerosol particles will be investigated.

DEFINITION OF TERMS

For something to be "seen" there must be a "source" to provide the radiant energy and a "receptor" to receive and translate this energy into an image. The source can be some type of radiant energy emitter such as the sun, a light bulb, or an aerosol particle (the carbon particles in a flame, for example), or it can be an object from which light from some other source is scattered — such as a building or a tree or an aerosol particle (in this case the smoke from a fire). The receptor is usually the eye, but it could be an instrument such as a photometer or a photomultiplier tube. It is not enough, however, to have a source and receptor, although these are necessary conditions. For something to be seen there must also be contrast between the object and its background. Without contrast the receptor cannot distinguish between the two.

Source represents the emission of radiant energy, which may be per unit time or area or both. In the cgs system of units, radiant energy is measured in ergs; in the mks system it is measured in joules. Other radiometric units evolve as listed in Table 14.1.

For visible light other definitions have evolved with the concept of a *luminator* as the source of luminous energy and *lumination* as the process. The definitions are shown in Table 14.2, arranged in a manner similar to Table 14.1. Units such as the lumerg are hardly ever used because the cgs system has fallen into disfavor among optical physicists, but are listed here because of the attempt within this book to stay with a consistent set of units throughout.

Fortunately most problems involving aerosol optics do not require the use of absolute quantities but instead make use of such unitless radiometric indicators as absorptance, reflectance, and the like. Table 14.3 lists several of the more common of these indicators. Figure 14.1 illustrates some of the more common definitions of light intensity.

In some works involving aerosol scattering the term albedo is used to describe the extinction of light by a particle or a system of particles. *Albedo* is defined as

Table 14.1. Radiometric (physical) concepts[a]

Name	Symbol	cgs. unit	mks. unit
Radiant energy	U	erg	joule
Radiant density	u	erg/cm^3	$joule/m^3$
Radiant flux	P	erg/sec	watt
Radiant emittance	W	$erg/(sec \times cm^2)$	$watt/m^2$
Radiant intensity	J	$erg/(sec \times \omega(*))$	$watt/\omega$
Radiance	N	$erg/(sec \times \omega \times cm^2)$	$watt/w \times m^2)$
Irradiance	H	$erg/(sec \times cm^2)$	$watt/m^2$
Spectral reflectance	r		
Spectral transmittance	t		

(*) ω = unit solid angle. The unit is normally the steridian.
[a]W. E. K. Middleton, *Vision Through the Atmosphere,* Univ. of Toronto Press, Toronto, 1963, p. 6.

Table 14.2. Psychophysical (photometric) concepts[a]

Name	Symbol	cgs unit	mks unit
Luminous energy	Q	lumerg	lumen-sec
Luminous density	q	$lumerg/cm^2$	$lumen-sec/m^2$
Luminous flux	F	lumerg/sec	lumen
Luminous emittance	L	$lumerg/(sec \times cm^2)$	$lumen/m^2$
Luminous intensity	I	$lumerg/(sec \times \omega)$	$lumen/\omega$ [candle]
Luminance	B	$lumerg/(sec \times \omega \times cm^2)$	$[candle/m^2]$
Illuminance	E	$lumerg/(sec \times cm^2)$	$lumen/m^2$ [lux]
Luminous reflectance	R		
Luminous transmittance	T		

[a]W. E. K. Middleton, *Vision Through the Atmosphere,* Univ. of Toronto Press, Toronto, 1963, p. 7.

Table 14.3. Unitless radiometric indicators[a]

Quantity	Symbol	Defining Equation	Unit
Absorbance	a	$a = (*)absorbed/(*)incident$	(numeric)
Reflectance	ρ	$\rho = (*)reflected/(*)incident$	(numeric)
Transmittance	τ	$\tau = (*)transmitted/(*)incident$	(numeric)

*Represents the appropriate quantity such as U, P, B, etc.
[a]D. Sliney and M. Wolbarsht, *Safety with Lasers and Other Optical Sources,* Plenum Press, New York, 1980, p. 937.

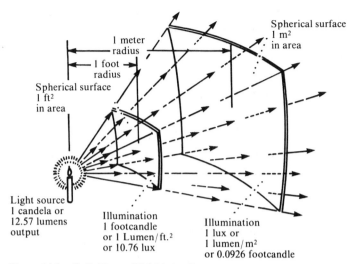

Figure 14.1. Definitions of light intensity.

the fraction of incident light or radiant energy that is reflected or scattered by the particle or system of particles. It is a dimensionless fraction commonly used to describe the light reflected from the earth back into space.

EXAMPLE 14.1

A satellite is used to relay television signals to the earth. If scattering and absorption by the earth's atmospheric aerosol causes the earth's albedo to be 0.60, how much of the signal from the satellite reaches the earth?

$$\text{fraction of signal reaching earth} = 1.00 - 0.6 = 0.4$$

A fundamental factor required for consideration of the optical properties of aerosols is the particle refractive index, m. The refractive index of a material is defined as the ratio of the speed of light in the material to the speed of light in a vacuum. Refractive indices of various materials are given in Table 14.3.

For aerosol particles where there is appreciable absorption of radiation as well as scattering it is necessary to express the refractive index of a material as a complex number of the form

$$m = v - ik \tag{14.1}$$

Here the parameter v represents the real part of the index and k the imaginary part. The real part represents scattering and the imaginary part absorption. Carbon particles, for example, have a refractive index of about $(2 - i)$. The materials listed in Table 14.4 are all reflective, absorbing little radiation. Thus their refractive indices are all expressed as real numbers.

Most aerosol materials will vary in their refractive index depending on the wavelength of light used, their chemical composition and, in some cases, their orientation with respect to the light source and receptor. Since complex indices of refraction are not well established for most materials (Deirmendjian, 1969), optical models of aerosols using these indices should be expected to contain errors because of the uncertainty of these values.

Having sorted out the various units which can be used (or at least having exposed the reader to them), it is now possible to consider optical properties of aerosols both within and outside of the range of visible wavelengths (0.4 μm to 0.7 μm.

Table 14.4. Refractive indices for selected materials

Material	m
Vacuum	1.000
Air	1.0002918
Water	1.3330
Diamond	2.417
Ice	1.31
Glass	1.48–2.00
NaCl	1.5443
PSL	1.5

EXTINCTION OF LIGHT – BOUGUER'S LAW

It has been established theoretically and experimentally that a thin section, dl, of a medium such as air which contains particles, will both scatter and absorb light in an amount proportional to the flux of light entering the section. For scattering, with F the luminous flux as defined in Table 14.2, and b a factor of proportionality,

$$dF = -bF\, dl \qquad (14.2)$$

Equation 14.2 integrates to

$$F = F_o e^{-bl} \qquad (14.3)$$

The path length from source to receptor is l. For absorption, with k as the factor of proportionality, an equation similar to Eq. (14.2) gives

$$F = F_o e^{-kl} \qquad (14.4)$$

The coefficients b and k are the scattering coefficient and absorption coefficient, respectively, and can be combined into a single extinction coefficient, ($\gamma = b + k$), so that

$$F = F_o e^{-(b+k)l} = F_o e^{-\gamma l} \tag{14.5}$$

The factor γ is also known as the turbidity or attenuation coefficient. Equations (14.3)–(14.5) are all forms of what is sometimes known as *Beer's law* but should more properly be called *Bouguer's law*, in honor of the person who empirically established it in 1760.

EXAMPLE 14.2

In a certain experiment the luminous flux, F, is reduced to 36.8% of its original value when a beam of light is passed through an aerosol over a path length of 10 meters. Determine the numerical value of γ in meters^{-1}.

From Eq. (14.5)

$$F = F_o e^{-(b+k)l} = F_o e^{-\gamma l} = 0.368$$

$$-\gamma l = ln(0.368)$$

$$-\gamma l = (-1.00)$$

$$\gamma = 0.100 \text{ m}^{-1}$$

With many particles of the same size,

$$\gamma = nQ_{ext} A \tag{14.6}$$

where n is the number of particles per unit volume of medium and Q_{ext} is an extinction efficiency factor defined as

$$Q_{ext} = \frac{\text{total energy flux extinguished by a single particle}}{\text{total energy flux geometrically incident on the particle}}$$

A scattering efficiency factor Q_{scat} and an absorption efficiency factor Q_{abs} can each be similarly defined. Then, from Eq. (14.5)

$$Q_{ext} = Q_{scat} + Q_{abs} \tag{14.7}$$

For spherical particles Eq. (14.6) becomes

$$\gamma = nQ_{ext} \frac{\pi}{4} d^2 \tag{14.8}$$

With a polydisperse cloud having n_i particles of cross-sectional area A_i, it is necessary to sum over all particles sizes, or,

$$\gamma = \sum_{i=1}^{\infty} n_i Q_{\text{ext }i} A_i \qquad (14.9)$$

This can be treated as an integral by considering $n(d)dd$ particles per unit volume having diameters in the interval d to $d + dd$. The total number of particles per unit volume is

$$n = \int_0^{\infty} n(d)dd \qquad (14.10)$$

For spheres, then, considering that Q_{ext} is also a function of d,

$$\gamma = \int_0^{\infty} \frac{\pi}{4} d^2 n(d) Q_{\text{ext}} \, dd \qquad (14.11)$$

Sometimes it is possible to choose appropriate values for A and Q_{ext} such that

$$\gamma = n \overline{Q_{\text{ext}}} \, \overline{A} \qquad (14.12a)$$

where the bars represent average values.

Similar equations can be written for Q_{scat} and Q_{abs}.

$$\gamma = n \overline{Q_{\text{scat}}} \, \overline{A} \qquad (14.12b)$$

$$\gamma = n \overline{Q_{\text{abs}}} \, \overline{A} \qquad (14.12c)$$

EXAMPLE 14.3

An aerosol contains an average of one million particles per cubic foot (1 mppcf), with the particle of average surface having a diameter of 3 μm. Assuming that $Q_{\text{ext}} = 2$, determine the value of γ for this aerosol, in m^{-1}.

Recalling Eq. (14.12a),

$$A = \text{area of average particle} = \frac{\pi d^2}{4} = \frac{\pi}{4} (3 \times 10^{-4})^2$$

$$A = 7.07 \times 10^{-8} \text{ cm}^2$$

$$n = 10^6 \text{ p/cf} = 3.53 \times 10^1 \text{ p/cm}^3$$

$$\lambda = (3.53 \times 10^1)(2)(7.07 \times 10^{-8})$$

$$\lambda = 5 \times 10^{-6} \text{ cm}^{-1} = 5 \times 10^{-4} \text{ m}^{-1}$$

For the case of a thin cloud consisting of spherical particles whose concentration is m g/cm^3, γ can be expressed in terms of the mass concentration by recalling that $m = (\pi/6)d^3\rho$ so that γ becomes

$$\gamma = nAQ_{ext} = \frac{3mQ_{ext}}{2\rho d} \tag{14.13}$$

When d is expressed in centimeters, γ has the units cm^{-1}. For a constant Q_{ext} and mass of material, Eq. (14.13) indicates that decreasing the particle size increases the extinction of light for an aerosol. That is, for the same amount of mass, small particles produce more haze in the atmosphere than large particles.

EXAMPLE 14.4

A typical urban aerosol has an average particle concentration of 75 μg/m^3. Assuming that the average particle can be represented by a 0.1-μm sphere with a density of 1 g/cm^3, determine the value of γ, in m^{-1}, assuming $Q_{ext} = 2$.

From Eq. 14.13

$$75 \ \mu g/m^3 = 75 \times 10^{-12} \ g/cm^3$$

$$\gamma = \frac{(3)(75 \times 10^{-12})(2)}{(2)(1)(10^{-5})} = 2.25 \times 10^{-5} \ cm^{-1}$$

$$\gamma = 2.25 \times 10^{-3} \ m^{-1}$$

Assumptions Implicit in Bouguer's Law

Utilization of Bouguer's law for extinction estimates assumes that certain simplifying assumptions are met. For example, it is assumed that the scattered light will have the same wavelength as the incident light. Although this is not precisely the case, wavelength changes are so small that they can be neglected.

A second assumption is that the particles act as independent scatterers, that is, the scattering of light by one particle does not influence the scattering by another. If particles are separated by more than about two diameters this assumption is met (Van deHulst, 1957). For 1-μ-diameter particles, this means concentrations in the order of 1.25×10^{11} particles/cm^3 before the assumption breaks down. At a concentration of this magnitude other factors such as coagulation (see Chapter 16) come into play to reduce the concentration, so this assumption is always valid.

A third assumption is that of single scattering, which is another way of saying that a maximum of one scatter per photon is allowed. This assumption implies that Bouguer's law is valid only for thin clouds or low concentrations, that is, the product of γl should not exceed 0.1 (Hodkinson, 1966). A simple way to test for the absence of multiple scattering is to observe if the scattering intensity is doubled when concentration is doubled. If so, single scattering prevails.

Computation of Extinction Coefficient

The term Q_{ext} represents an efficiency factor. In Examples (14.3) and (14.4) it was assumed Q_{ext} had a constant value of 2. In actuality Q_{ext} is a function of d, the particle diameter. For large particles Q_{ext} is equal to two. For very small particles, $Q_{abs} \approx 0$ so that $Q_{ext} \approx Q_{scat}$. Then

$$Q_{scat} = Q_{ext} = \frac{8}{3} \alpha^4 \left[\frac{m^2 - 1}{m^2 + 2} \right]^2 \tag{14.14}$$

where m is the refractive index of the particle. The particle diameter and wavelength of the incident light, λ, are related through the dimensionless factor, α, defined as

$$\alpha = \frac{\pi d}{\lambda} \tag{14.15}$$

Equation 14.14 applies when $\alpha \ll 1$. This special case of light scattering is called *Rayleigh scattering,* and one important example of Rayleigh scattering is the scattering of light by molecules making up the earth's atmosphere.

EXAMPLE 14.5

Compute the value of Q_{scat} for an "air" molecule when illuminated with blue light ($\lambda = 0.4\ \mu m$). Then compare this value to Q_{scat} for the same molecule illuminated with red light ($\lambda = 0.7\ \mu m$). Assume a molecular diameter of 4×10^{-8} cm and $m = 1.000292$.

$$\alpha_{blue} = \frac{\pi(4 \times 10^{-8})}{0.4 \times 10^{-4}} = 3.14 \times 10^{-3}$$

$$\alpha_{red} = \frac{\pi(4 \times 10^{-8})}{0.7 \times 10^{-4}} = 1.80 \times 10^{-3}$$

$$Q_{scat\ blue} = \frac{8}{3}(3.14 \times 10^{-3})^4 \left[\frac{(1.000292)^2 - 1}{(1.000292)^2 + 2} \right]^2$$

$$Q_{scat\ blue} = 9.82 \times 10^{-18}$$

$$Q_{scat\ red} = \frac{8}{3}(1.80 \times 10^{-3})^4(3.79 \times 10^{-8})$$

$$Q_{scat\ red} = 1.06 \times 10^{-18}$$

Notice that in this case blue light is scattered about nine times as efficiently as red light.

For larger particles the relationship of Q_{ext} to d is more complex, initially rising according to the fourth power relationship of Rayleigh scattering until it finally oscillates around and then becomes asymptotic to a value of $Q_{ext} = 2$. For water droplets, a plot of Q_{ext} versus α is shown in Fig. 14.2. Oscillations in the value of Q_{ext} are due to internally reflected light being in or out of phase during scattering.

The asymptotic value of $Q_{ext} = 2$ implies that a particle can remove light from an area equal to twice its cross-section. This so-called extinction paradox

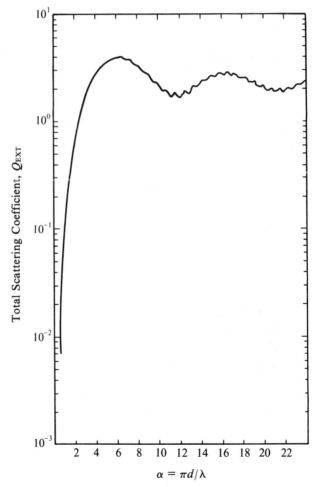

Figure 14.2. Plot of Q_{ext} versus α for water, $m = 1.33$.

arises from light being diffracted by the particle into the space behind it from which other photons had previously been scattered or absorbed. This amount will just be equal to that diffracted through a hole having the same cross-sectional area as the particle. As a result the total amount of energy removed from the proceeding wave will be twice that shadowed out by the particle itself. Hence $Q_{ext} = 2$.

The asymptotic value of $Q_{ext} = 2$ is not only important for particles which are primarily scatterers but also for irregularly shaped particles or those which are highly absorbing. These types of particles reach the value of $Q_{ext} = 2$ at fairly small particle sizes. For example, for an irregular, transparent particle $Q_{ext} = 2$ is reached when $\alpha(m - 1) > 10$, or for a salt crystal, when its edge is approximately 3 μm.

By experimental measurement of light extinction it is possible to estimate a value for Q_{ext} when absorption is considered. Experiments by Connor and Hodkinson (1967) indicate that for absorbing particles Q_{ext} rises rapidly from 0 to a value of 2 without the oscillations which are so evident for transparent spherical particles. The final constant value of $Q_{ext} = 2$ is reached when $\alpha \approx 0.5$ to 1. This can be seen in Table 14.5 which lists values of Q_{ext} for white smoke ($m = 1.5$) and black smoke ($m = 1 - 0.5i$) illuminated with light having a wavelength of 0.5 μm.

Table 14.5. Extinction coefficients for absorbing particles

	Particle Diameter						
	0.1	0.2	0.4	0.8	1.6	3.2	6.4
Q_{ext} (white)	0.03	0.14	0.88	1.87	2	2	2
Q_{ext} (black)	–	0.88	1.87	2	2	2	2

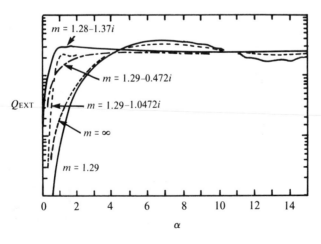

Figure 14.3. Q_{ext} as a function of α for aerosol particles of various refractive indices. (D. Deirmendjian, *Electromagnetic Scattering on Spherical Polydispersions,* Elsevier, New York, 1969, p. 28.)

The effect of absorption on the value of Q_{ext} can also be seen from the values shown in Fig. 14.3 (Deirmendjian, 1969). In these calculations the real part of m was kept constant with only the imaginary part being varied. The maximum value of Q_{ext} that could be reached is reduced by absorption and the efficiency of smaller particles is increased. A totally reflecting sphere ($m = \infty$) is not as efficient over some size ranges in removing light as particles which have some absorption.

EXAMPLE 14.6

Two aerosols, both consisting of 0.4-μm-diameter particles, contain equal particle concentrations. One aerosol is made up of white particles, the other of black particles. How much greater will be the reduction in luminous flux for the black smoke compared to the white smoke? The path length for measurement is the same in both cases.

$$(\text{ratio}) = \frac{F/F_0 \,_{\text{black}}}{F/F_0 \,_{\text{white}}} = \frac{e\gamma_b l}{e\gamma_w l}$$

$$\ln(\text{ratio}) = \frac{\gamma_b l}{\gamma_w l} = \frac{\gamma_b}{\gamma_w}$$

$$\ln(\text{ratio}) = \frac{nQ_{ext}(\text{black})A}{nQ_{ext}(\text{white})A}$$

From Table 14.5

$$\ln(\text{ratio}) = \frac{1.87}{0.88} = 2.13$$

$$\text{Ratio} = e^{2.13} = 8.37 \text{ times greater}$$

RECEPTOR–CONTRAST

To see an object, whether it is an emitter or not, requires a certain level of contrast between the object and its background. Without contrast the object blends into its background and becomes invisible. For an isolated object surrounded by a uniform and extensive background contrast can be defined as

$$C = \frac{B}{B'} - 1 \tag{14.16}$$

where B is the luminance of an object and B' the luminance of the background. When these are equal the contrast, C, is equal to zero, indicating that the object cannot be seen. When an object is less luminous than its background C has a negative value reaching -1 for a black object against a white background. For the opposite case C can go to infinity. A light at night represents a large positive C value.

The question of how much contrast is necessary in order for an object to be seen is one of much importance and much speculation. From extensive studies carried out during World War II it appears that this contrast level depends on such tangibles as the medium through which the object is being viewed (is the object being viewed through air or water?) and also on intangibles such as the psychological state of the viewer, the adaption of his eyes, and so on. For daylight conditions with a black object being viewed against a white background, a value of $C = -0.02$ is often used. Although rough, this value, known as the threshold of brightness contrast, can be used as a guide for visibility approximations.

EXAMPLE 14.7

Determine the value of B/B', which gives rise to $C = -0.02$.

From Eq. 14.6

$$C = \frac{B}{B'} - 1 = -0.02$$

$$\frac{B}{B'} = 1 - 0.02 = 0.98$$

The luminance of the object is 98% of the luminance of the background.

ALTERATION OF CONTRAST

Aerosols in the atmosphere change the contrast of the atmosphere. This is evident to anyone who has viewed objects at a distance. Visibility decreases as the atmospheric load of aerosol material increases.

The first attempt to explain the alteration of contrast was made by Koschmieder (1924a,b). In developing his model for the alteration of contrast he determined the amount of light from around a black object scattered into the vision of an observer such that the observer would think the scattered light came from the object. To do this it was necessary for Koschmieder to make a number of simplifying assumptions. He assumed that the atmosphere contains a large number of small particles and each volume element contains a large number of particles, much smaller than the element. He further assumed that each volume element can be treated as a point source with independent scattering, a cloudless sky (equal illumination to all parts of the atmosphere in a horizontal plane), a constant scattering coefficient, b, the earth's curvature can be neglected, the observed object is small compared to the distance of the object to the observer and refraction of light by the earth's atmosphere is negligible.

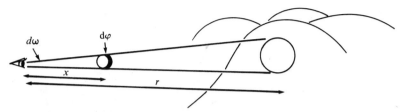

Figure 14.4. Model conditions for Koschmieder derivation.

A volume element is then defined as shown in Fig. 14.4 where the element has the volume

$$d\varphi = d\omega x^2 dx \qquad (14.17)$$

Since this volume element will be illuminated exactly as the object, regardless of x, the luminous intensity scattered from the volume element toward the observer is

$$dI = Abd\varphi \qquad (14.18)$$

Where A is a constant for all values of x. The illuminance at the eye of the observer due to the light from $d\varphi$ is

$$dE = dI x^{-2} e^{-bx} \qquad (14.19)$$

The x^2 term arises from the inverse square law, indicating the amount of light coming from $d\varphi$ at a particular distance x to the observer relative to all light coming from that distance x. The e^{-bx} term represents the scattering of light along the path from $d\varphi$ to the eye.

$$dB = \frac{dE}{d\omega} = Abe^{-bx} dx \qquad (14.20)$$

$$B = \int_0^r Abe^{-bx} dx = B_h(1 - e^{-br}) \qquad (14.21)$$

B_h is the illumince of the horizon sky.
From Eq. (14.16)

$$C = \frac{B}{B_h} - 1 = -e^{-br} \qquad (14.22)$$

Next ϵ is defined as the threshold of brightness contrast. This represents the contrast level where a black object can just be seen against a white background. Using $\epsilon = -0.02$, recognizing that this value can vary by as much as an order of magnitude (see previous section), then

$$-0.02 = -e^{-br} \qquad (14.23)$$

Rewriting in terms of r gives

$$r = -\frac{1}{b} \ln 0.02 = \frac{3.91}{b} \tag{14.24}$$

The term r is known as the *visual range* and has units of length. Since in actuality concentration can vary over the distance between the object and the observer, r should more properly be called the *meteorological range* (Charlson et al., 1967).

EXAMPLE 14.8

What is the estimated visual range in a monodisperse aerosol having a concentration of 0.1 mg/m^3 if the particle diameter is 0.1 μm and $Q_{scat} = 2$? Assume $\rho = 1$ gm/cm^3.

From Eq. 14.12b

$$\gamma = n \overline{Q_{scat} \, A}$$

$$A = \frac{\pi}{4} d^2 = \frac{(3.14)(10^{-5})^2}{4} = 7.85 \times 10^{-11} \text{ cm}^2$$

$$n = \frac{6m}{\pi d^3 \rho} = \frac{(6)(0.1 \times 10^{-3}/10^6)}{(3.14)(10^{-15})(1)} = 1.91 \times 10^5 \text{ p/cm}^3$$

$$b = (1.91 \times 10^5)(2)(7.85 \times 10^{-11}) = 3.00 \times 10^{-5} \text{ cm}^{-1}$$

$$r = \frac{3.91}{3 \times 10^{-5}} = 1.30 \times 10^5 \text{ cm} = 1.3 \text{km}$$

It might be thought that the visual range equation could be used to measure aerosol mass concentration in the atmosphere since visual range is a fairly simple measurement. Indeed, some studies comparing predicted mass to measured mass concentrations tend to bear this assumption out. The difficulty lies in choosing a proper average particle diameter, d. As pointed out in Chapter 12, the average size of atmospheric aerosol particles can vary markedly, depending on their moisture content. For soluble nuclei this can be further confounded by the hysteresis effect, by which the value of d will be determined by whether the nuclei are in an atmosphere of rising or falling humidity. Since very often this fact is difficult to ascertain, especially with a moving air mass measured at a stationary point, mass concentration measurements derived from extinction measurements should only be considered to be valid for cases where the atmospheric humidity is less than 40%.

PROBLEMS

1. If the luminance of the sky near the horizon on a moonless clear night sky is 10^{-3} candles/m^2, determine whether a star of magnitude 5 (0.1 candles/m^2) would be seen.

2. Charlson, Horvath and Pueschel (1967) give the empirical expression

$$\text{mass } (\mu g/m^3) = 3 \times 10^5 b$$

where b is in meters. Assuming $Q_{scat} = 2$, what particle size does this relationship imply?

3. Using the expression in problem 2, show that

$$l \times \text{concentration} = 1.2 \text{ g/m}^2$$

What is the physical significance of this product?

4. Determine Q_{scat} for a 0.01-μm and 0.05-μm diameter water droplet using blue light.

5. The particulate standard relating visual range to particulate concentration is

$$R = \frac{120}{G}$$

where R is the range in kilometers and G is the particle concentration in micrograms per cubic meter. Assuming that Q_{ext} equals 2, what average particle diameter is implied by this standard?

CHAPTER FIFTEEN

OPTICAL PROPERTIES
Angular Scattering

DEFINITIONS

The terms Q_{scat}, Q_{abs}, and Q_{ext} represent loss of radiation along the path from source to observer, that is, the extinction of light by an aerosol. But often interest centers more on the scattering of light in a single direction. The diameter of a particle can be estimated by the quantity of light scattered from the particle into a detector. Or, if a thundercloud is to be tracked by radar, the intensity of radiation backscattering is important.

To establish a frame of reference, θ is defined as the forward angle of scattering as shown in Fig. 15.1, representing the angle between the direction of propagation of the light and the direction of scattering. An angle of $\theta = 0°$ represents "forward scattering", i.e., light scattered along the direction of propagation.

A scattering intensity coefficient $q(\theta)$ can be defined as

$$q(\theta) = \frac{\text{flux scattered into } \theta}{\text{flux geometrically incident on particle}} \tag{15.1}$$

The incident light is considered to be parallel and may or may not be polarized. If it is polarized, then the scattered light must also be polarized in the same plane. If the incident light is not polarized, the scattered light may or may not be polarized. With polarized light $q_1(\theta)$ is defined as the scattering intensity co-

232

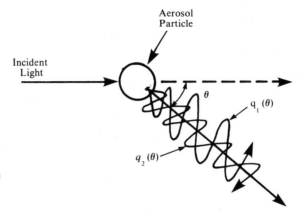

Figure 15.1. Sketch of definitions $q_1(\theta)$ and $q_2(\theta)$ for directional scattering from an aerosol particle.

efficient for light propagated in a plane normal to the plane formed by the incident and scattered light vectors; $q_2(\theta)$ is the scattering intensity coefficient for light propagated parallel to that plane.

For unpolarized incident light

$$q(\theta) = \frac{1}{2}[q_1(\theta) + q_2(\theta)] \qquad (15.2)$$

The $q_1(\theta)$ or $q_2(\theta)$ components can be observed by polarizing either the incident or scattered radiation. Then, if I_0 is the intensity of an incident beam of perpendicularly polarized light, the intensity of the beam scattered per unit solid angle by a particle of diameter d will be

$$I = q_1(\theta) I_0 \frac{\pi}{4} d^2 \qquad (15.3)$$

A similar expression can be written for plane polarized light. For n monodisperse particles per unit volume, assuming single scattering,

$$I = n q_1(\theta) I_0 \frac{\pi}{4} d^2 \qquad (15.4)$$

In terms of $q_1(\theta)$ and $q_2(\theta)$, the relationship for the total scattering efficiency of a particle is

$$Q_{\text{scat}} = \pi \int_0^\pi [q_1(\theta) + q_2(\theta)] \sin \theta \, d\theta \qquad (15.5)$$

EXAMPLE 15.1

For Rayleigh scattering the perpendicularly polarized light is scattered equally in all directions, $q_1(\theta) = c$; the plane polarized light is scattered according to $q_2(\theta) =$

$c \cos^2 \theta$, where the constant c depends on the values of α and m. If c is equal to $\alpha^4/\pi[(m^2 - 1)/(m^2 + 2)]^2$, use Eq. (15.5) to determine Q_{scat} for Rayleigh scattering.

$$Q_{scat} = \pi \int_0^\pi (c + c \cos^2 \theta) \sin \theta \, d\theta$$

$$Q_{scat} = 2\pi c \left(-\cos \theta - \frac{\cos^3 \theta}{3} \right)_0^{\pi/2}$$

$$Q_{scat} = 2\pi c \left[\frac{4}{3} \right] = \frac{8}{3} \alpha^4 \left[\frac{m^2 - 1}{m^2 + 2} \right]^2$$

MIE SCATTERING – THE MIE THEORY

Determination of the value of $q(\theta)$ as a function of m and θ for all values of α and the two polarization states can be accomplished through the use of the "Mie Theory of Radiation Scattering." Unfortunately solutions of Mie's equations do not lend themselves readily to numerical computation. Manageable solutions other than Mie theory are available for the cases where $\lambda > d$ (Rayleigh scattering) or when $\alpha \gg 1$ (geometric optics). For the intermediate region, representing particle diameters from about 0.1 µm to 10 µm, Mie solutions must be used.

Although Mie's theory was first published in 1908, computations of scattering coefficients were not tabulated to any extent until the 1940s (Lowan, 1948), and then the available tables were quite limited. Nevertheless, considering that each data point represented many hours of error-free calculation with a desk calculator, the accuracy of these early tables is indeed remarkable.

With the advent of high-speed electronic computers calculation of Mie scattering functions has become routine, if not commonplace, and library programs are available for this task. Kerker (1969) gives a listing of various sources of computed Mie functions and discusses problems associated with the mechanics of computation.

Mie scattering functions are generally presented in terms of the intensity parameters for Mie scattering, also known as the angular intensity functions, $i_1(\theta)$ and $i_2(\theta)$. The subscripts of these functions indicate perpendicular and plane polarization respectively. Besides being functions of the scattering angle θ, $i_1(\theta)$ and $i_2(\theta)$ are also functions of the particle properties m and α (e.g. Lowan [1948] or Denman, Heller, and Pangonis [1966]).

For a given α and m the angular intensity functions are related to the scattered intensity coefficients, $q_1(\theta)$ and $q_2(\theta)$, by the expressions

$$q_1(\theta) = \frac{i_1(\theta)}{\pi \alpha^2} \tag{15.6}$$

$$q_2(\theta) = \frac{i_2(\theta)}{\pi \alpha^2} \tag{15.7}$$

Plots of $i_1(\theta)$ and $i_2(\theta)$ as a function of θ for $m = 1.33$ and three values of α are given in Figs. 15.2a, 15.2b, and 15.2c. Figures 15.3a, 15.3b, and 15.3c are plots of $i_1(\theta)/\alpha^3$ as a function of α for $0°$, $90°$, and $180°$ scattering respectively. Although representing only one refractive index, that of water droplets, these figures show the variations in $i_1(\theta)$ and $i_2(\theta)$ expected for many transparent aerosol materials. It can be seen that as α increases, the forms of both $i_1(\theta)$ and $i_2(\theta)$ become more complex.

The degree of polarization of light scattered from a particle that is illuminated with unpolarized light can be estimated from the ratio $[i_1(\theta) - i_2(\theta)] / [i_1(\theta) + i_2(\theta)]$;

EXAMPLE 15.2

Perpendicularly polarized white light is used to illuminate a 0.5-μm-diameter sphere. If the sphere has a refractive index of 1.33, what will be the predominant color of light scattered at an angle of $90°$?

For visible light and a 0.5-μm-diameter sphere, alpha can range in value from 2.24 to 3.93. From Fig. 15.3b, $i_1(\theta)$ is a maximum within this range when alpha is equal to 2.8. This represents a light wavelength of 0.56 μm, representing a color of violet-blue.

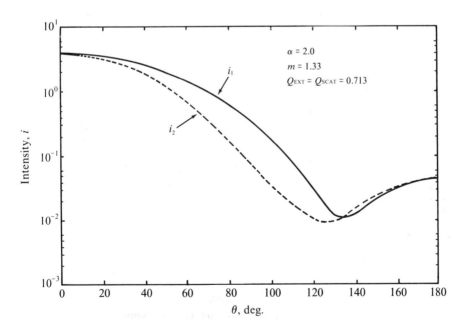

Figure 15.2a. Angular intensity functions, $i_1(\theta)$ and $i_2(\theta)$, for a sphere; $\alpha = 2.0$, $m = 1.33$.

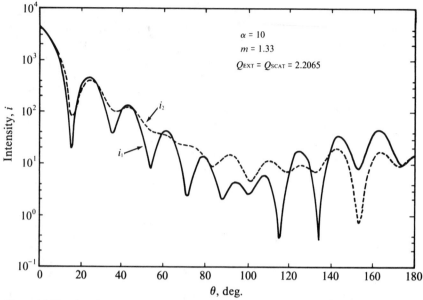

Figure 15.2b. Angular intensity functions, $i_1(\theta)$ and $i_2(\theta)$, for a sphere; $\alpha = 10$, $m = 1.33$.

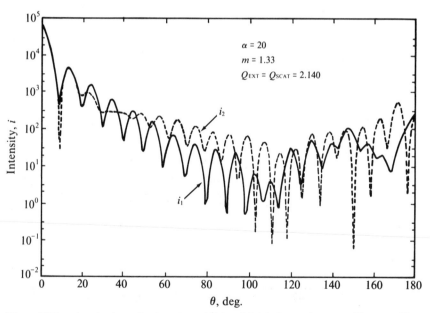

Figure 15.2c. Angular intensity functions, $i_1(\theta)$ and $i_2(\theta)$, for a sphere; $\alpha = 20$, $m = 1.33$.

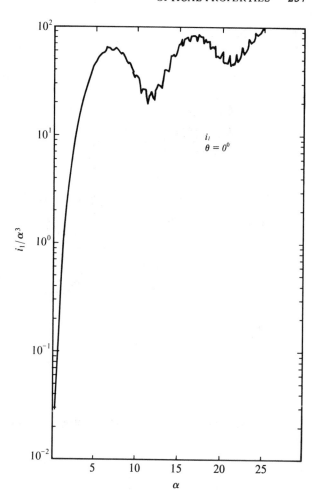

Figure 15.3a. Plot of i_1/α^3 for 0° scattering angle, $m = 1.33$.

Approximations to Mie Theory

There have been many attempts to find approximations for the Mie extinction and intensity functions that would be more wieldy, but for the most part the search has been unsuccessful. Penndorf [1962] gives an approximation for $i_1(\theta)$ for forward scattering ($\theta = 0°$) which he claims to be valid for $\alpha > 5$ and any m. This equation is

$$i_1(0°) \approx \left(\frac{\alpha^2}{4} Q_{\text{scat}}\right)^2 \tag{15.8}$$

Table 15.1 compares some values of $i_1(0°)$ Calculated from Eq. 15.8 with values from more exact Mie theory computations.

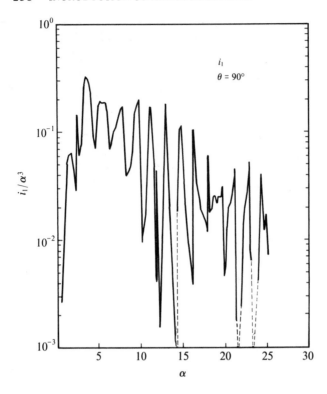

Figure 15.3b. Plot of i_1/α^3 as a function of α, $m = 1.33$.

Table 15.1. Comparison of values of $i_1(0°)$ calculated from Eq. 15.8 and Mie theory[a]

α	Q_{scat}	m	$i_1(0°)$ from Mie Theory	$i_1(0°)$ from Eq. 15.8
5	3.928	1.5	610.1	602.7
10	2.882	1.5	5209	5191
15	1.985	1.5	12,490	12,500
20	2.036	1.5	42,570	41,500
25	2.250	1.5	127,500	124,000
30	2.353	1.5	282,400	280,000

[a]1.5 refractive index data from R. Giese et al., *Tabellen der Streufunktionen* $i_1(0)$, $i_2(0)$, Akademie-Verlag, Berlin, 1962, p. 10.

EXAMPLE 15.3

Compute the value of $i_1(0°)$ for a 3-μm-diameter sphere ($m = 1.5$) when illuminated with blue light ($\lambda = 0.4\mu$m). Assume $Q_{scat} = 2$.

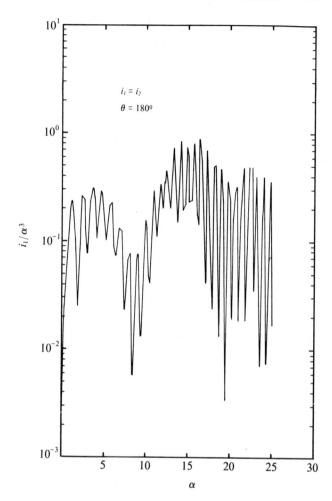

Figure 15.3c. Plot of $i_1/\alpha^3 = i_2/\alpha^3$ versus α for spheres. $m = 1.33$, $\theta = 180$.

$$\alpha = 23.56$$

$$i_1(0°) = \left(\frac{(23.56)^2(2)}{4}\right)^2$$

$$i_1(0°) = 7.70 \times 10^4$$

For $\alpha = 23.6$ and $m = 1.5$ Giese et al. (1962) give for $i_1(0°)$ a value of 1.27×10^5:

According to Kerker (1969) it is possible to estimate the $i_1(\theta)$ values for scattering in the near forward direction (scattering angles no greater than several degrees) over the range $\alpha = 5$ to 30 using the approximation

$$i_1(\theta) \approx i_1(0°)\left[\frac{J_1(\alpha \sin \theta)}{\alpha \sin \theta}\right]^2 \tag{15.9}$$

In Eq. 15.9 J_1 represents a Bessel function of the first order and of argument $(\alpha \sin \theta)$. Appendix G lists tabulated values of positive Bessel functions of order 1.

The value of Q_{scat} can be estimated using Van de Hulst's approximation (1957) for transparent spheres,

$$Q_{scat} = 2 - \frac{4}{\psi} \sin \psi + \frac{4}{\psi^2}[1 - \cos \psi] \tag{15.10}$$

The term ψ is defined as $\psi = 2\alpha(m - 1)$, and in Eq. (15.10), ψ is expressed in radians.

EXAMPLE 15.4

Denman et al. (1966) give a computed value for $i_1(5°)$ as 1.91×10^4 for a 4-μm diameter particle ($m = 1.33$) illuminated with light having a wavelength of 0.5 μm. Compare this value to one calculated using Eq. (15.10).

$\alpha = 25.1$

$\psi = (2)(25.1)(1.33 - 1) = 16.59$

$Q_{scat} = 2 - [4/16.59] \sin (16.59) + [4/(16.59)^2][1 - \cos (16.59)] = 2.21$

$i_1(0°) = \left(\frac{(25.1)^2(2.21)}{4}\right)^2 = 1.22 \times 10^5$

$\alpha \sin \theta = 2.188$

From Appendix G, $J_1(2.188) \approx 0.556$

$i_1(5°) = 1.22 \times 10^5(0.556/2.19)^2 = 7.84 \times 10^3$

This result is somewhat lower than that given by the reference above.

Polydisperse Aerosol

With scattering from a volume containing a polydisperse aerosol both the total scattering cross-section and scattering functions show much less irregularity as maxima and minima are smoothed out by the variety of particle sizes present. According to computations by Foitzik et al. (1965/1966), irregularities in an

$i_1(\theta)$ versus θ curve are essentially gone for log-normally distributed aerosols having a $\sigma_g > 1.4$ (Fig. 15.4).

Rayleigh Scattering

In the case where the particle size is less than the wavelength of light, $\alpha < 0.3$, the electromagnetic field can be assumed to be uniform over the entire particle, giving equations for scattering that are relatively simple. This approach was successfully used by Lord Rayleigh (1899) to explain the scattering of light by "air" molecules. To honor this finding it is common practice today to refer to the scattering of light when α is very small as *Rayleigh scattering* or scattering in the *Rayleigh region*.

Rayleigh was able to explain that natural light scattered at right angles to its source is completely polarized. He reasoned that a photon of natural light approaching a particle along the *x*-axis (centered at the origin of Cartesian co-

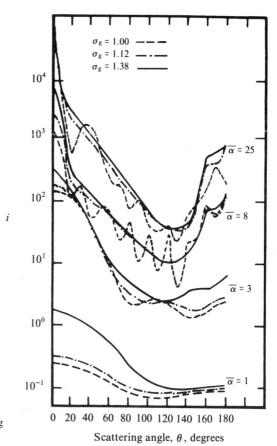

Figure 15.4. Integrated scattering function *i*.

Scattering angle, θ, degrees

ordinates) can be resolved into waves vibrating in the y- and z-directions. At the particle this photon produces a dipole oscillating in the same directions. Since there is no forward or x-component of wave motion for photons scattered into an angle of $90°$ to the incident beam, there will be only one component of the wave motion, that being the component perpendicular to the plane formed by the incident and scattered beams. That is, the $i_1(\theta)$ component will remain and the $i_2(\theta)$ component will disappear. By the definitions given above, $q_2(90°) = 0$ for scattering in the Rayleigh region and the scattered beam is perpendicularly polarized (see for example Kerker, 1969, or Van de Hulst, 1957).

Rayleigh made the important discovery that the intensity of light scattered by very small particles is proportional to the fourth power of the wavelength of the incident light. This finding explains the blue color of the sky on a clear day, as well as the apparent blue color of smokes or fumes made up of very small particles.

For very small nonabsorbing spheres Rayleigh's theory gives

$$i_1(\theta) = \alpha^6 \left[\frac{m^2 - 1}{m^2 + 2}\right]^2 \tag{15.11}$$

$$i_2(\theta) = \alpha^6 \left[\frac{m^2 - 1}{m^2 + 2}\right]^2 \cos^2 \theta \tag{15.12}$$

Thus for a particle illuminated with natural incident light of intensity I_0, the intensity per unit solid angle is

$$I = \frac{(1 + \cos^2 \theta)}{2} I_0 \left[\frac{m^2 - 1}{m^2 + 2}\right]^2 \frac{\alpha^6}{k^2} \tag{15.13}$$

where k is the wave number defined as $k = 2\pi/\lambda$. From the ratio of a α^6/k^2 the fourth power dependence of scattering on wavelength is shown.

EXAMPLE 15.5

Compute the intensity of natural light scattered at a $45°$ angle by a 0.01-μm water droplet ($m = 1.33$) when illuminated by

a. red light ($\lambda = 0.7 \ \mu m$)
b. blue light ($\lambda = 0.4 \mu m$).

$$\left[\frac{\alpha^6}{k^2}\right]_{red} = 1.01 \times 10^{-18}$$

$$I = \frac{(1 + 0.500)}{2} I_0 \left[\frac{1.33^2 - 1}{1.33^2 + 2}\right]^2 1.01 \times 10^{-10}$$

$$I/I_{0 \ red} = (0.750)(0.042)(1.01 \times 10^{-18}) = 3.15 \times 10^{-20}$$

$$\left[\frac{\alpha^6}{k^2}\right]_{blue} = 9.51 \times 10^{-18}$$

$$I = \frac{(1 + 0.500)}{2} I_0 \left[\frac{1.33^2 - 1}{1.33^2 + 2}\right]^2 9.51 \times 10^{-10}$$

$$I/I_{0\text{ blue}} = (0.750)(0.042)(9.51 \times 10^{-18}) = 2.97 \times 10^{-19}$$

The relationships between the values $i_1(\theta)$ and $i_2(\theta)$ for Rayleigh scattering are sketched in polar coordinates in Fig. 15.5. The factor $i_1(\theta)$ forms a circle centered at the origin while $i_2(\theta)$ produces two circles tangent at the origin and lying on the $\theta = 0°$ line. The disappearance of the $i_2(\theta)$ scattering at $\theta = 90°$ can be seen.

Scattering Patterns with Increasing α

As the particle size parameter becomes larger the backscattered component slowly begins to weaken and the forward scattering increases in intensity. With the increasing assymetry comes the development of maxima and minima which, with increasing α, develop toward the rear of the scattering pattern and then move toward the front, constantly growing. Eventually, for monodisperse aerosols, these maxima and minima form complex scattering patterns having as their most outstanding characteristics irregularity and a strong forward scattering component. The development of a complex scattering pattern with increasing α is illustrated in Fig. 15.6.

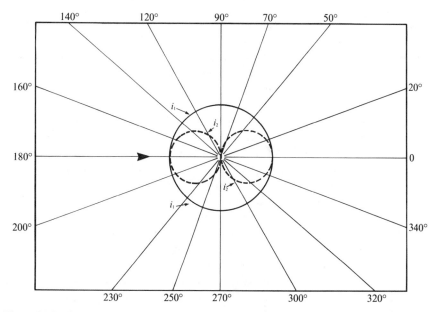

Figure 15.5. Polar diagram of angular scattering functions, $i_1(\theta)$ and $i_2(\theta)$ for Raleigh scattering; $\alpha \ll 1$, $m = 1.33$.

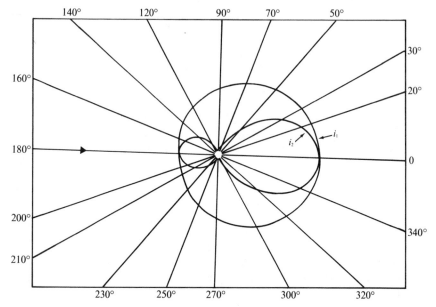

Figure 15.6. Polar diagram of angular scattering functions, $i_1(\theta)$ and $i_2(\theta)$ for $\alpha = 1.0$, $m = 1.33$.

Thus for aerosols in the sizes of interest, particle scattering is characterized by the strong forward scattering of both plane polarized and perpendicularly polarized components. For nonspherical or irregularly shaped particles of the same size range this strong forward scattering tendency is also evident, although there are fewer maxima and minima in the angular scattering pattern. For aerosol particles in the range of interest there is also a pronounced backscatter (180°) component. As mentioned previously, the degree of irregularity in the scattering curves decreases as the aerosol becomes more polydisperse and the forward scattering component increases.

Smaller sizes of particles (and for a polydisperse cloud, more particles) will be seen when the particles are viewed between the light source and the observer than when viewed with side illumination. This observation can be confirmed easily by comparing the number of dust particles seen by looking up a shaft of sunlight with particles viewed at right angles to the sun.

Forward scattering of light by dust particles in the atmosphere is responsible for the red color of the sun as it sets, and also for the glorious sunsets that herald evening. Just as the sun is setting, when cloud conditions are right, it is possible to observe a red glow in the east, as well as the sunset in the west. In this case the 180° backscattered light is responsible.

The scattering diagrams for particles in the Mie region show many maxima and minima, some quite close together, but slightly altered in angular position with small changes in α. These anomalies are responsible for the strange colors that can appear in the sky or in an experimental chamber from time to time when conditions of particle size and monodispersity are right. For example, Aitken (1892) reported that on some rare occasions the sun appeared greenish; other times it appeared bluish. Following a very large forest fire in Canada in the early 1950s both the sun and moon appeared bluish for several days. This led Cadle (1966) to remark that the rare occurrence of such a cloud of particles in the atmosphere could be the source of the expression "once in a blue moon."

RADIATIVE TRANSFER

If multiple interactions between photons and aerosol particles do take place, the resulting model of this behaviour is extremely complicated. This class of problem is often described as the "problem of radiative transfer," similar to the neutron scattering problem. In concept the approach is quite simple. For example, in Fig. 15.7 various aerosol particles are shown as dots. These particles may be similar or dissimilar. Light radiating from Plane A can reach Plane B either by single scattering, as shown by scattering from particle R, or by multiple scattering, shown by scattering from particle P to particle Q and thence to Plane B. Any number of scattering events are possible before the light photon reaches Plane B, more scatterings being likely as the aerosol concentration increases.

When all possible photon paths from A are considered and all photons reaching B are summed, the problem is solved. Various approximations have been proposed to reduce this almost infinite task, but a complete solution hs not yet been developed. For more information on this mathematically interesting problem the reader is referred to Kerker (1963).

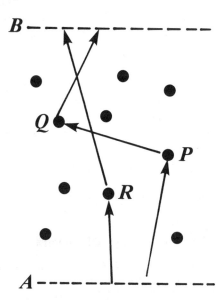

Figure 15.7. Model of multiple scattering.

APPLICATIONS

Particle size represents an important variable in both the extinction and scattering of light. It is not surprising, therefore, that there is a wide variety of optical techniques available for particle sizing. In this section several of the more common methods will be discussed.

Diffraction Rings

If a light source is viewed through a thin, fairly monodisperse aerosol, a series of bright rings (coronas) will be seen. If d is the diameter of the cloud particles and γ_n the angular radius of the nth minimum between the rings, then the following empirical relationships are valid:

$$\sin \gamma_1 = \frac{1.22\lambda}{d} \tag{15.14a}$$

$$\sin \gamma_2 = \frac{2.24\lambda}{d} \tag{15.14b}$$

$$\sin \gamma_3 = \frac{3.24\lambda}{d} \tag{15.14c}$$

When using white light (assuming $\lambda = 0.571$ μm), the outer edge of the ring is considered to be the minimum. Often only the light scattered from the first ring is intense enough to be seen. This phenomenon can be seen around the moon on some nights prior to the arrival of a cold front. If the moon is used as the light source, its angular radius should be subtracted from the total angular radius before using Eqs. 15.14a, b, or c. This technique has been used to derive particle size information in very high clouds such as altostratus clouds.

EXAMPLE 15.6

On a cloudy night the first diffraction ring around a blue light source ($\lambda = 0.4$ μm) having an apparent angular radius of $7°$ appeared at an angle of $15°$. What is the particle diameter implied by these measurements?

$$\sin (15 - 7) = 1.22(0.4)/(d)$$

$$d = 3.51 \ \mu\text{m}$$

Higher Order Tyndall Spectra

If one illuminates a monodisperse aerosol with white light, a series of well-defined colors are produced at various scattering angles. These colors appear because of the extreme dependence of the scattered intensity at a given angle on the value of α, and hence, for a monodisperse aerosol, on the wavelength of light. As a result, at a given angle one particular wavelength will scatter much more light than any of the other visible wavelengths, and one color will predominate. Johnson and LaMer (1947) found that over a fairly wide range of refractive indices the number and angular positions of a given color (red was chosen because it was considered easiest to see) was a function of particle size. For a traverse of $180°$ the relationship can be approximated by:

$$d \approx \text{number of reds from } 0° \text{ to } 180°/5 \qquad (15.15)$$

More exact computations of this relationship, based on Mie theory, are shown in Figs. 15.8 and 15.9. A device known as the *aerosol owl* can be used to make these measurements. It consists of a viewing chamber on which a telescope is mounted that can be rotated through roughly $160°$. A protractor is attached to the telescope so that the angle at which a red is observed can be noted. The number of reds found can be used to determine particle size, which can also be confirmed by comparing the angles at which the reds appear with Fig. 15.9. A more complete description of the owl is given by Dennis (1976) or Green and Lane (1964), and discussions of the use of higher order Tyndall spectra are presented by Kitani (1960) and Maron and Elder (1963).

EXAMPLE 15.7

A monodisperse aerosol is viewed in an aerosol owl and four red bands are observed as the sample is scanned over approximately $160°$. Determine the aerosol particle diameter indicated by these data.

From Eq. (15.15)

$$d \approx \frac{4}{5} \approx 0.8 \, \mu\text{m}$$

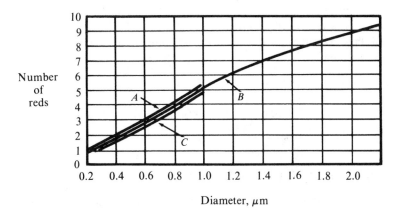

Figure 15.8. Observed and theoretical number of reds. A = sulfur ($m = 2.0$), B = stearic acid ($m = 1.43$), C = calculated (all indices).

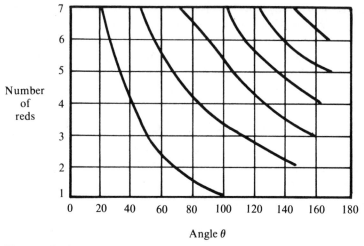

Figure 15.9. Angular position of reds for stearic acid aerosol.

Use of the Forward Scattering Lobe

A technique for sizing monodisperse particles of unknown refractive index uses the observation that the intensity of the forward scattering lobe is mainly due to Fraunhofer diffraction, and is thus independent of the refractive index of the particle. Hodkinson (1966) computed a family of curves for the ratio of intensity viewed at two different angles. This relationship is shown in Fig. 15.10. If forward angles as small as $5°$ can be measured, this procedure should be accurate in measuring particle sizes of $\alpha = 1$ to $\alpha = 18$.

EXAMPLE 15.8

The ratio of the intensity of light scattered at an angle of $15°$ to light scattered at an angle of $10°$ when a monodisperse aerosol is illuminated with unpolarized red light ($\lambda = 0.5$ μm) is 0.35. What particle diameter is implied by this measurement?

From Fig. 15.10, $\alpha = 12.7$.

$$d = \frac{\alpha\lambda}{\pi} = (12.7)(0.5)/(3.24) = 2.02 \ \mu m$$

Single Particle Scattering Measurements

The scattering technique for particle sizing that has gained the most popular acceptance was first developed by Gucker and O'Konski (1949) and since then

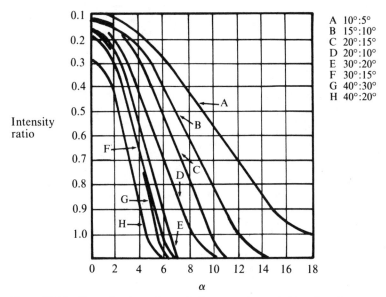

Intensity ratio

A 10°:5°
B 15°:10°
C 20°:15°
D 20°:10°
E 30°:20°
F 30°:15°
G 40°:30°
H 40°:20°

Figure 15.10. Scattering intensity ratio for scattering at indicated angles, as a function of α.

has been developed to such an extent that now there are many commercially available instruments using this technique. Particle size is determined by passing the particles, one at a time, through an illuminated volume, and the amount of light scattered by each particle into some solid angle is measured. Knowledge of the flow rate and the number of particles counted per unit time gives the aerosol concentration, whereas the amount of light scattered per particle gives an indication of that particle's size. This type of instrument is, in theory, capable of measuring both the concentration and size distribution of an aerosol. A good discussion of these types of aerosol detectors has been given by Whitby and Willeke (1979).

Single-particle optical counters cannot be easily used for measuring high aerosol concentrations. Whitby and Liu (1967) have shown that if many small particles are in the illuminated sensitive volume of this type of instrument at the same time, significant errors are introduced because the particles are considered by the detector to be fewer in number and greater in size. Thus there is a practical upper limit on the aerosol concentration that can be measured by such an instrument, being in the order of 1000 to 10,000 particles/cm^3.

EXAMPLE 15.9

An optical particle counter samples an aerosol at a flowrate of 175 cm^3/min into a sensitive volume of 0.01 cm^3. What is the implied maximum particle concentration (in particles per liter) than can be sampled?

To have a maximum of one particle in the sensitive volume at any time, on the average there should be no more than 100 particles/cm^3, or 100,000 particles/liter.

Because of the statistical distribution of particles entering the counter, even though an average of only one particle will be in the chamber at any time, actually sometimes there will be more than one particle, other times no particles at all. Hence the theoretical limit given in Example 15.9 cannot be achieved without some coincidence losses. The sensitive volume of the optical counter could be considered to require a certain time, t_R, to recover from counting one particle prior to counting another. This is the average time taken for a particle to pass through the sensitive volume,

$$t_R = \text{(volume of sensitive volume)/(volumetric flow rate)}$$

The observed counting rate can be increased to account for coincidence losses by adding to the counting rate a factor Θ given by

$$\Theta = \frac{C^2 t_R}{1 - C t_R} \qquad (15.16)$$

EXAMPLE 15.10

Using the data in Example 15.9 determine the corrected aerosol concentration when the indicated concentration is 23,100 particles/liter.

In 1 min, 23,100 (particles/liter) $\times$ 0.175 (liters/min) particles are counted,

$$= 4{,}040 \text{ counts/min}$$

$$t_R = (0.01 \text{ cm}^3)/(175 \text{ cm}^3/\text{min}) = 5.71 \times 10^{-5} \text{ min}$$

$$\Theta = \frac{(4.04 \times 10^3)^2 (5.71 \times 10^{-5})}{[1 - (4.04 \times 10^3)(5.71 \times 10^{-5})]} = 1210 \text{ additional cpm}$$

Thus corrected counting rate is 4040 + 1,210 = 5250 cpm.
or corrected concentration is

$$= \frac{5250}{0.175} = 30{,}000 \text{ particles/liter.}$$

Rearrangement of Eq. (15.16) shows that a saturation count will be reached by the counter where increasing particle challenge does not increase particle count. Since this saturation level is fairly low for most optical particle counters compared to typical aerosol concentrations, for field use some sort of dilution system is usually required.

PROBLEMS

1. The terms $i_1(\theta)$ and $i_2(\theta)$ are related to $q_1(\theta)$ and $q_2(\theta)$, the scattered intensity coefficients, by the equations

$$i_1(\theta) = q_1(\theta) \left(\frac{\pi}{4}\right) d^2 k^2$$

$$i_2(\theta) = q_2(\theta) \left(\frac{\pi}{4}\right) d^2 k^2$$

where k is the wave number, $k = 2\pi/\lambda$. If, for Rayleigh scattering

$$q_2(\theta) = \frac{\alpha^4}{\pi} \left[\frac{m^2 - 1}{m^2 - 1}\right]^2 \cos^2 \theta$$

plot on polar coordinate paper $q_1(\theta) + q_2(\theta)$.

2. The value of $q_1(\theta)$ for perpendicularly polarized light scattered in the Rayleigh region is given by

$$q_1(\theta) = \frac{\alpha^4}{\pi} [(m^2 - 1)/(m^2 + 2)]^2$$

Determine the scattering efficiency for light scattered into the interval $\theta = 0°$ to $20°$.

3. Perpendicularly polarized white light is used to illuminate a 0.25-μm-diameter sphere. If the sphere has a refractive index of 1.33, what will be the predominant color of light scattered at an angle of $90°$?

4. Using Eq. (15.8), compute the value of $i_2(0°)$ for a 1-μm-diameter sphere when illuminated with light having a wavelength of 0.5 μm.

5. Infrared rays have wavelengths in the range 0.7 to 5000 μm. For 10-μm-diameter particles, what values of α correspond to the wavelengths of the infrared rays?

6. A 1.0-μm-diameter particle is illuminated with visible light having a wavelength of 0.5 μm and infrared radiation having a wavelength of 10 μm. Assuming equal intensity of the incident radiation, find the ratio of the scattered visible to infrared radiation for forward scattering.

SIXTEEN

COAGULATION OF PARTICLES

INTRODUCTION

The aerosol properties discussed in previous chapters relate primarily to individual particles. For the most part, discussions have avoided consideration of interference effects between particles. But one area where interparticle effects cannot be neglected is aerosol coagulation, also known as aggregation or agglomeration.

It is easily observed that at high aerosol concentrations individual particles coalesce to form larger chains or flocs made up of many particles. The process of coagulation may be brought about solely by the random motion and subsequent collision of particles (often called thermal coagulation) or the collisions could be caused by such external forces as turbulence or electricity. In general these external forces will act to increase the rate of coagulation.

The process of coagulation has received a great deal of attention, especially with regard to the coagulation of colloidal solutions. Unfortunately, because particle number as well as particle size changes with time, numerical models that predict these variables as a function of time are highly complex and therefore messy to use, or, if simplified, are representative but inexact. In many cases results derived from the simplified models can be used to predict aerosol coagulation rates and number concentration with reasonable accuracy, although little can be said of the resulting size distributions.

COAGULATION OF MONODISPERSE SPHERICAL PARTICLES

The simplest coagulation problem is thermal coagulation of monodisperse spherical particles. Since only the first several particle collisions are considered, the size of the resulting agglomerated particles will not be appreciably different from the size of the initial particles. The model has been used for many years for solid particles and forms the basis for the definition of the coagulation coefficient. It is especially applicable to the coagulation of liquid drops, since the size of the agglomerated drop increases only as the cube root of the number of drops making it up. The approach was first presented by Smoluchowski (1911) for coagulation in dilute electrolytes, but it was shown to be applicable also to aerosols, within the limits discussed above (Whytlaw-Gray and Patterson, 1932).

In Smoluchowski's approach a number of spherical particles of diameter d are considered to be randomly separated from each other at $t = 0$. If the particles are also moving about randomly by thermal diffusion, it is desired to know the likelihood that they will collide (and then stick together) within some time period, t. Smoluchowski first considered the case where one particle was fixed, acting as a sink for the other particles. He then determined the diffusion rate of other particles to this central particle.

For diffusion

$$\frac{\partial c}{\partial t} = D\nabla^2 c \tag{16.1}$$

where c is the concentration of particles and D is their diffusion coefficient. If r is a distance measured from the center of the fixed particle, and assuming spherical symmetry, Eq. (16.1) can be written as

$$\frac{\partial c}{\partial t} = D\left(\frac{\partial^2 c}{\partial r^2} + \frac{2}{r}\frac{\partial c}{\partial r}\right) \tag{16.2}$$

or in an even more convenient form, as

$$\frac{\partial(cr)}{\partial t} = D\frac{\partial^2(cr)}{\partial r^2} \tag{16.3}$$

EXAMPLE 16.1

Show that Eq. (16.3) is equivalent to Eq. (16.2).

If c is a function of t and r,

$$\frac{\partial(cr)}{\partial t} = r\frac{\partial c}{\partial t}$$

and

$$\frac{\partial(cr)}{\partial r} = r\frac{\partial c}{\partial r} + c$$

differentiating with respect to r gives

$$\frac{\partial^2(cr)}{\partial r^2} = r\frac{\partial^2 c}{\partial r^2} + \frac{\partial c}{\partial r} + \frac{\partial c}{\partial r}$$

Then, substituting in Eq. (16.3) gives

$$r\frac{\partial c}{\partial t} = D\left(2\frac{\partial c}{\partial r} + r\frac{\partial^2 c}{\partial r^2}\right)$$

or

$$\frac{\partial c}{\partial t} = D\left(\frac{\partial^2 c}{\partial r^2} + \frac{2}{r}\frac{\partial c}{\partial r}\right)$$

Since all particles are the same diameter, it can be assumed that they will intersect the central particle when they come to within a distance of d from it. Thus at this point the concentration will be zero, i.e.,

$$c' = 0 \text{ at } r = d (\text{for } t > 0).$$

In addition, initially particles are assumed to be uniformly distributed throughout the volume of interest with a concentration of c. Thus

$$c' = c \text{ at } t = 0$$

With these conditions, Eq. (16.3) can be solved to give

$$c' = c\left[1 - \frac{d}{r} + \frac{d}{r}\operatorname{erf}\left(\frac{r-d}{2\sqrt{Dt}}\right)\right] \tag{16.4}$$

The number of particles, Φ, which diffuse to within a distance d of the central fixed particle in unit time is equal to the product of the diffusion current and surface area of the sphere of radius d. The diffusion current is given by the expression

$$J = -D \frac{\partial c'}{\partial r} \tag{16.5}$$

where the slope, $\partial c'/\partial r$, is to be evaluated at $r = d$. Thus

$$\Phi = 4\pi d^2 D \frac{\partial c'}{\partial r} \tag{16.6}$$

Since at $r = d$, from Eq. (16.4),

$$\frac{\partial c'}{\partial r} = \frac{c}{d} \left[1 + \frac{d}{\sqrt{\pi D t}} \right] \tag{16.7}$$

then in the time interval, dt, the number of particles reaching the surface surrounding the central particle is

$$\Phi dt = 4\pi d D c \left[1 + \frac{d}{\sqrt{\pi D t}} \right] dt \tag{16.8}$$

Now suppose that the fixed particle is not fixed, but is able to diffuse along with the other particles. The diffusion of this particle must also be taken into account. The combined diffusion coefficient of two particles relative to each other is equal to the sum of the diffusion coefficients of the single particles so that the moving particle collides with

$$8\pi d D c \left[1 + \frac{d}{\sqrt{\pi D t}} \right] dt$$

particles in dt, remembering that Eq. (16.8) applies to equal-sized spheres. Also, with c particles per unit volume there will be $c/2$ collisions if every particle collides once, there being two particles involved in each collision. The number of collisions per unit volume which will take place in the time interval dt is

$$\frac{dc}{dt} = -\frac{8}{2}\pi d D c^2 \left[1 + \frac{d}{\sqrt{\pi D t}} \right] \tag{16.9}$$

The second term in the brackets can be ignored, since it will generally be much less than one, especially after some time has elapsed. This can be seen by referring to Table 16.1 which gives values of the term $d/\sqrt{\pi D} = \xi$ for various size particles. Fuchs (1964) points out that ξ represents the possibility of particles initially being quite close to the fixed particle. Thus ξ disappears as a stationary rate develops. Since for all practical cases ξ is quite small, it is ignored. One should keep in mind, however, that there could be conditions where ξ would be of importance. The effect of this term, if any, is to increase the coagulation rate.

Table 16.1. Calculated values of ξ

Particle diameter, cm	$d/\sqrt{\pi D} = \xi$
1×10^{-7}	2.42×10^{-7}
5×10^{-7}	6.01×10^{-6}
1×10^{-6}	2.39×10^{-5}
5×10^{-6}	5.67×10^{-4}
1×10^{-5}	2.12×10^{-3}
5×10^{-5}	3.54×10^{-2}
1×10^{-4}	0.107
5×10^{-4}	1.28
1×10^{-3}	3.65

Defining the coagulation constant, K_0, as

$$K_0 = 8\pi dD = \frac{8}{3}\frac{kT}{\mu}C_c \tag{16.10}$$

yields a results which should be, at least for large particles, independent of particle size.

EXAMPLE 16.2

Determine the value for the coagulation constant, K_o, for air at 20°C, standard pressure for particles having $C_c \approx 1$.

$$K_0 = \frac{(8)(1.38 \times 10^{-16})(293)(1)}{(3)(1.83 \times 10^{-4})}$$

$$K_0 = 5.89 \times 10^{-10} \text{ cm}^3/\text{sec}$$

In the cgs system the units of K_0 are cubic centimeters per second.

Using K_0 as the coagulation constant and neglecting the second term in Eq. (16.9) gives the usual form of the coagulation equation:

$$\frac{dc}{dt} = -\frac{K_0}{2}c^2 \tag{16.11}$$

Integration of Eq. (16.11) with the initial condition that $c = c_0$ when $t = 0$, gives

$$\frac{1}{c} - \frac{1}{c_0} = \frac{K_0}{2}t \tag{16.12}$$

Equation (16.12) shows that the inverse of the concentration at any time is a linear function of the time, the slope of the line being determined by the coagulation constant. Experimental data from both monodisperse and polydisperse aerosols follow this general form, at least initially. As will be discussed later, the coagulation constant may be appreciably larger than the theoretical value. If t_h is defined as the half-value time, that is, the time in which the concentration decreases by a factor of two, then

$$c = \frac{c_0}{1 + (t/t_h)} \qquad (16.13)$$

EXAMPLE 16.3

An aerosol made up of 5-μm diameter spheres has an initial concentration of 100 million particles per cubic foot (100 mppcf). After one day what will be the aerosol concentration (also in mppcf)? How long will it take to reach half the initial concentration?

$$K_0 = 5.89 \times 10^{-10} \text{ cm}^3/\text{sec}$$

$$100 \text{ mppcf} = (100)(10^6)/(28.3)(10^3) = 3.53 \times 10^3 \text{ p/cc}$$

$$\frac{1}{c} - \frac{1}{3.53 \times 10^3} = \frac{5.89 \times 10^{-10}}{2}(60 \times 60 \times 24)$$

$$c = 3.24 \times 10^3 \text{ p/cc} = 91.8 \text{ mppcf}$$

To find the time to reach half the initial concentration,

$$\frac{2}{c_0} - \frac{1}{c_0} = \frac{K_0}{2} t_h$$

$$t_h = \frac{2}{c_0 K_0} = \frac{2}{(3.53 \times 10^3)(5.89 \times 10^{-10})}$$

$$t_h = 9.62 \times 10^5 \text{ sec} = 11.13 \text{ days}$$

COAGULATION OF PARTICLES OF TWO DIFFERENT SIZES

For the case of coagulation of particles of two different sizes it is possible to follow the same approach as given for the monodisperse case except that d is replaced with $(d_1 + d_2)/2$ and $2D$ is replaced with $(D_1 + D_2)$. Then the coagulation coefficient K_{12} becomes

$$K_{12} = 2\pi(d_1 + d_2)(D_1 + D_2) \qquad (16.14)$$

In terms of the particle mobility the coagulation coefficient is

$$K_{12} = 2\pi(d_1 + d_2)(B_1 + B_2)kT \tag{16.15}$$

As previously defined, B, the mobility, is

$$B = \frac{D}{kT} = \frac{C_c}{3\pi\mu d}$$

The minimum coagulation constant occurs for coagulation of equal-sized particles. This can be seen in Fig. 16.1 which shows a three-dimensional graph of the coagulation constant matrix. The valley indicated on the plot represents the constants for coagulation of equal-sized particles.

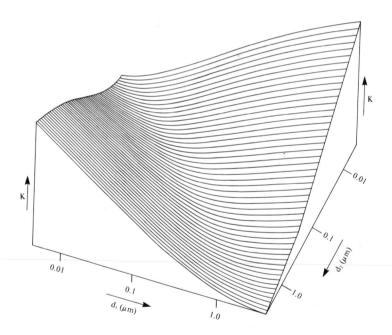

Figure 16.1. Coagulation constant, K, matrix, symmetric in d_1, d_2 (from Kuhlman, 1982 Minimum value is 4.4×10^{-10} cm^3/sec. Maximum value is 1.28×10^{-5} cm^3/sec.

EXAMPLE 16.4

Compute the combined coagulation coefficient for coagulation of 1.0-μm and 0.1-μm -diameter spheres.

From Table 8.1, D for a 1-μm particle = 2.74 $\times$ 10^{-7} cm^2/sec. D for a 0.1-μm particle = 6.82 $\times$ 10^{-6} cm^2/sec.

$$K_{12} = 2\pi(10^{-4} + 10^{-5})(2.74 \times 10^{-7} + 6.86 \times 10^{-6})$$

$$K_{12} = 4.90 \times 10^{-9} \text{ cm}^3/\text{sec}$$

Coagulation of Many Sizes of Particles

Tikhomirov, Tunitskii, and Petryanov (1942) as well as Gillespie (1963) have shown that if one is interested only in the change in the number of all the particles, dc/dt, then it is possible to combine all coagulation coefficients into a single one that will be expressed in terms of the mean values $\bar{d}$, $\overline{(1/d)}$ and $\overline{(1/d^2)}$. This combined coefficient, K', is

$$K' = \frac{4kT}{3\mu}\left[1 + \bar{d}\overline{\left(\frac{1}{d}\right)} + A\lambda\overline{\left(\frac{1}{d}\right)} + A\lambda\bar{d}\overline{\left(\frac{1}{d^2}\right)}\right] \qquad (16.16)$$

where A is a Cunningham correction factor constant equal to 1.26, and λ is the mean free path of the gas. Thus K' can be used as before in Eq. (16.11).

If particle sizes are initially distributed log-normally, an expression can be written for K' in terms of the geometric mean diameter d_g and geometric standard deviation, σ_g, that is,

$$K' = \frac{4kT}{3\mu}\left[1 + \exp(\ln^2\sigma_g) + \frac{2A\lambda}{d_g}\exp\left(\frac{1}{2}\ln^2\sigma_g\right) + \frac{2A\lambda}{d_g}\exp\left(\frac{5}{2}\ln^2\sigma_g\right)\right] \qquad (16.17)$$

Table 16.2 lists values of K' for various d_g and σ_g. With increasing polydispersity this coefficient can become quite large, implying that extremes in polydispersity are quickly reduced by agglomeration, particularly of the smaller particles. Also, it is clear that the coagulation rate for a polydisperse aerosol is greater than for a monodisperse one. However, coagulation of a monodisperse aerosol initially increases polydispersity, so that for any coagulating aerosol the coagulation coefficient is not constant, but is itself a variable bound to the coagulation rate. There is little wonder that the interpretation of coagulation data is difficult, and it is amazing that the simple coagulation theory, as presented here, is as adequate as it is for many applications.

Table 16.2. Values of the coagulation coefficient K' for aerosols of varying degrees of polydispersity [value $\times 10^{-10}$ cm^3 sec^{-1}]

Geometric mean diameter d_g, μm	1.0	1.5	Geometric standard deviation, σ_g 2.0	2.5
0.02	53.2	67.8	116.4	238.8
0.04	29.6	37.1	62.1	124.3
0.10	15.4	18.7	29.5	55.6
0.16	11.8	14.1	21.3	38.4
0.20	10.6	12.6	18.6	32.7
0.40	8.23	9.49	13.1	21.2
1.00	6.84	7.65	9.88	14.3
1.60	6.48	7.19	9.07	12.6
2.0	6.37	7.03	8.79	12.1
4.0	6.13	6.74	8.25	10.9
10.0	5.99	6.54	7.93	10.2
16.0	5.95	6.50	7.85	10.1

Computed with the following constants, $\rho = 1$ g/cm^3, $k = 1.38 \times 10^{-16}$, $T = 293°C$, $\mu = 1.83 \times 10^{-4}$ poises, $\lambda = 0.0653\mu$m.

EXAMPLE 16.5

Figure 16.2 shows a plot of $(1/c)$ as a function of t for a polydisperse aerosol. Using this figure determine the value of K'.

From the figure the slope of the line can be found by

$$\text{slope} = \frac{\frac{1}{c_2} - \frac{1}{c_1}}{t_2 - t_1}$$

$$\text{slope} = \frac{3.92 \times 10^{-8} - 0.68 \times 10^{-8}}{85 - 10}$$

$$= 4.32 \times 10^{-10}$$

$$K' = 2 \times 4.32 \times 10^{-10} = 8.64 \times 10^{-10} \text{ cm}^3/\text{sec}$$

DIFFERENTIAL EQUATION FORM

Suppose there exists an aerosol which initially consists of particles of sizes 1, 2, 3, ... up to size m. The change in concentration of the kth-size particles results not only from a loss of particles from this size group by combination with other particles, but also from a gain in particles into this size group by the right combination of smaller particles. This can be written as

$$\Omega = \Psi - \Phi \tag{16.17}$$

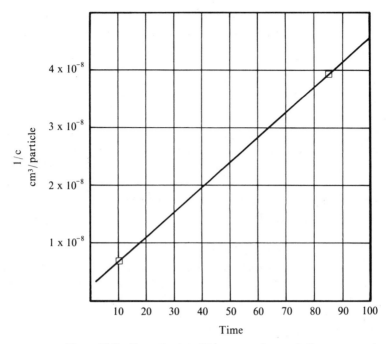

Figure 16.2. Example plot of $1/c$ versus t for a polydisperse aerosol.

where Ω = change in number in the kth-size, Ψ = increase due to combination making a kth-size particle, and Φ = loss of kth-size particle by combination, or, more formally,

$$\frac{dc_k}{dt} = \frac{1}{2} \sum_{j=1}^{k-1} K_{j(k-j)} c_j c_{(k-j)} - \sum_{j=1}^{\infty} K_{kj} c_k c_j \qquad (16.18)$$

EXAMPLE 16.6

Given an aerosol made up of particles of discrete sizes $d_1, d_2, d_3 \ldots d_n$, write an equation expressing the change in the concentration of size d_4 particles with time, if no particles are lost from this size interval.

With no particle loss, the second term of Eq. (16.18) is zero. Then

$$\frac{dc_4}{dt} = \frac{1}{2} \left[K_{13} c_1 c_3 + K_{22} c_2 c_2 \right]$$

The total change in the number of particles of all sizes is equal to the sum of the change of the numbers of particles in the individual sizes, which can be written as

$$\frac{dN}{dt} = -\frac{1}{2} \sum_{k=1}^{\infty} \sum_{j=1}^{\infty} K_{kj} c_k c_j \tag{16.19}$$

which is just one-half the sum of the second part of Eq. (16.18) since there is no overall increase in the number of particles during coagulation. For Eq. (16.19) no sources or sinks have been assumed for the coagulating particles.

LIMITATIONS OF THE DIFFERENTIAL EQUATION FORM

In Eq. (16.18) the coagulation rates of particles of different sizes are represented by a coupled set of nonlinear ordinary differential equations. As coagulation proceeds, the number of equations that are required to describe the size distribution spectrum of an aerosol increases, so that in the extreme one might be required to solve a set of 1000 or more of these equations simultaneously. For example, to determine changes in size distribution of several coagulating monodisperse and polydisperse aerosols, Hidy (1965) solved up to 600 equations simultaneously and even then there were instances where material was "lost" from the system because it coagulated into sizes larger than the largest size allowed for the particles. As the calculations proceeded, this "loss" limited the accuracy of his numerical solutions.

USE OF A NONLINEAR INTEGRO-DIFFERENTIAL EQUATION

To circumvent the problem of many equations it is possible to represent the coagulation equation as a nonlinear integro-differential equation of the form

$$\frac{\partial(v, t)}{\partial t} = \frac{1}{2} \int_0^v K(\phi, v - \phi) c(\phi, t) c(v - \phi, t) d\phi$$
$$- \int_0^{\infty} K(\phi, v) c(\phi, t) c(v, t) d\phi \tag{16.20}$$

Particle volume is used rather than particle diameter because the volumes are additive, whereas the diameters are not. Otherwise, this equation is analogous to the set of ordinary differential equations — in one case a discrete model is used, in the other, a continuous model is chosen.

The first term in Eq. (16.20) represents the increase in particles in the volume size range $v + dv$ from the combination of particles of volume ϕ and volume

$(v - \phi)$. The term $c(\phi, t)$ represents the number of particles of volume ϕ at time t, $c(v - \phi, t)$ the number of particles of volume $(v - \phi)$ at time t, and the term $K(\phi, v - \phi)$ the coefficient for coagulation of particles of volume ϕ and volume $(v - \phi)$.

The second term of this equation represents the loss of particles in the volume size range $v + dv$ resulting from coagulation of particles of volume v and volume ϕ. The term $K(v, \phi)$ is the combined coagulation coefficient for these two particles. Thus this equation gives an expression for the net rate of change of particles whose volumes lie between v and $v + dv$. Solution of the equation with appropriate initial conditions gives the number of particles of volume $v + dv$ at any time t.

To determine the total number of particles per unit volume it is necessary to integrate over all particle volumes. Symbolically this is

$$N(t) = \int_0^\infty c(v, t) \, dv \tag{16.21}$$

Then the change in the total number of particles with time becomes

$$\frac{dN(t)}{dt} = -\frac{1}{2} \int_0^\infty \int_0^\infty K(v, \phi) c(v, t) c(\phi, t) \, dv \, d\phi \tag{16.22}$$

For the case of constant K and a homogeneous aerosol, Eq. (16.22) reduces to Eq. (16.11).

TERMS FOR GRAVITY AND DEPOSITION EFFECTS

An advantage to using Eq. (16.20) is that other mechanisms that may also remove particles from the coagulating aerosol cloud can be included. For example, removal of particles by gravity settling can be accounted for by adding the term

$$[-\vec{v}(v) \, \nabla \, n(v, t)]$$

to the right-hand side of Eq. (16.22); diffusional deposition onto boundary surfaces by the term

$$[D(v)\nabla^2 \, n(v,t)]$$

($\vec{v}$ is the velocity vector of a particle of volume v and D is its diffusional coefficient) (Takahashi and Kasahara, 1968). Thus a generalized equation can be formed which accounts for most, if not all, aerosol behavior with time.

THE "SELF-PRESERVING" SIZE DISTRIBUTION

It has been suggested by Friedlander (1965), Friedlander and Wang (1966), and Wang and Friedlander (1967), that a coagulating aerosol should with time reach the same steady-state size distribution regardless of the aerosol's initial size distribution. This distribution is called the *self-preserving size distribution.* When this steady state is reached, gains by coagulation in the number of particles of a given size are equaled by losses from that size either by coagulation or by sedimentation. For very small particles, sedimentation losses are unimportant; for very large ones coagulation losses can be neglected. This implies three distinct distribution functions over the entire particle size spectrum. However, with no allowance for a source of particles, these distribution functions are really quasi-steady state, since eventually the whole system will run out of particles making a null distribution function.

Hidy (1965) utilized a set of differential equations similar to Eq. (16.18) to test for the existence of self-preservation and found that spectral curves did indeed tend toward an asymptotic value with time. Later Hidy and Brock (1970) showed that value was itself a function of time and eventually disappears. Hidy's study determined that the quasi-steady-state spectrum would be fully developed in a time given by

$$t = \frac{9\mu}{kTc_o} \qquad (16.23)$$

Clark and Whitby (1967) used Friedlander's self-preserving spectrum theory to explain the general shape of the observed size distribution of atmospheric aerosols. Although the formation of the distribution is so slow that there is little likelihood of finding this form in most cases, it is possible that for some global aerosols the quasi-self preserving spectrum of Friedlander is actually developed.

COAGULATION OF NONSPHERICAL PARTICLES

For nonspherical particles, Muller (1928) postulated that since the diffusion equation applicable to aerosol problems is the same (except for definition of terms) as the general equation for electrical fields (LaPlace's equation), there should be analogs among the electrostatic terms for various properties of coagulation. For example, the potential should be analogous to particle number concentration, and field strength to particle agglomeration rate. Zebel (1966) points out that since field strengths are known to be high at points of sharp curvature, the particle agglomeration rate should also be high at similar places on an irregular particle. The analogy can be carried further to imply that shapes other than spherical tend to enhance coagulation, compared to large spherical particles. However, since particle mobility must also be considered and the decreased mobility of irregularly shaped particles would tend to depress coagulation, Muller's approach should only be regarded as qualitative, giving at best an idea of the effect of particle shape on coagulation rates. It does appear to be true that particle deposition takes place preferentially on sharp protuberances.

EXTERNAL FACTORS IN COAGULATION

A number of external physical factors can act to enhance or retard the coagulation of aerosols. These include electrical effects such as the attraction or mutual repulsion of charged particles, polarization effects giving rise to induced forces, sonic agglomeration, gravitational coagulation, and coagulation brought about by turbulence. In the following sections these effects are discussed only briefly since, in general, a detailed treatment is beyond the scope of this text. For more information the reader is referred to the appropriate references.

Electrical Effects in Coagulation

In coagulation in an electric field particles diffuse toward one another until they are close enough for electric forces to come into play. Then the motion of the particles assumes an ordered nature.

With electrically charged particles of the same or opposite sign coagulation may be enhanced or diminished depending on the signs of the two charges and their magnitude. A derivation for the ratio of the coagulation constants for charged and uncharged particles was first given by Fuchs (1934) and is discussed in detail in more recent references by Fuchs (1964) and Zebel (1966). A plot showing the results of Fuchs' derivation is given in Fig. 16.3. When $y < 0$ the

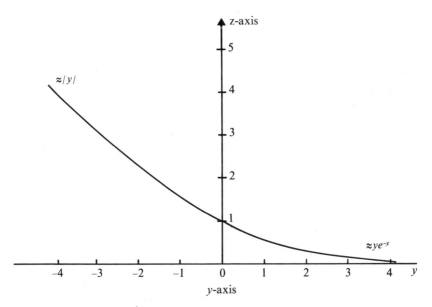

Figure 16.3. Plot of $Z = y/e^y - 1$.

forces between the particles are attractive, whereas when $y > 0$ they repel. For a weak bipolar aerosol the ratio z can be approximated by a straight line through the point $z = 1$ at $y = 0$, indicating that the increase in coagulation brought about by attraction is compensated for by a similar decrease due to repulsion (Zebel, 1966).

On the other hand, with a very strong bipolar aerosol $|y| \gg 1$, the increase in coagulation due to attraction greatly exceeds the decrease due to repulsion, giving a net increase in the coagulation rate.

For a unipolar aerosol it is necessary to consider electrostatic dispersion, that is, the tendency of charged particles of the same sign to move away from each other. This dispersion tends to reduce the concentration of an aerosol, for example by causing particles to deposit on the walls of any containing vessel or nearby surface, and thus can interfere with the measurement of electrical effects in coagulation. Hidy and Brock (1970) have used the Debye-Huckel model to analyze electrostatic effects on coagulation. They show that when electrostatic dispersion is considered strong, bipolar aerosols will have enhanced coagulation coefficients whereas strongly charged unipolar aerosols will have greatly reduced coagulation coefficients. They caution that these estimates are only approximate since polarization of the electric field can greatly alter the effect of charging in coagulation. Fuchs (1966) points out that the coagulation of mists is enhanced only in very strong electric fields (something in excess of 200 V/cm). For solid particles polarization in an electric field enhances the formation of chain-like aggregates. Particles that are permanent dipoles also form these chain-like structures.

Coagulation in Moving Atmospheres

Up to this point only Brownian motion has been considered. Coagulation can also take place in rapidly moving airstreams, and one might expect that by moving the particles coagulation would be enhanced. This motion could assume two forms. On the one hand there could be an ordered flow of particles in one direction, with the particles moving at different velocities. This could occur, for example, in a polydisperse aerosol settling by gravity under quiescent conditions. On the other hand, motion could be disordered, as would be the case with turbulent mixing.

COAGULATION WITH ORDERED MOTION

First consider the case of a large particle moving through a cloud of much smaller particles. What will be the decrease in number concentration of the smaller particles brought about by the larger one? Assuming that the ordered velocity of the smaller particles is negligible compared to the velocity of the larger one, it can be assumed that the large particle removes in unit time all smaller particles contained in a volume given by $(\pi/4)d_L^2 v_L$. The term d is the diameter of the larger particle and v_L its velocity. However, as discussed in Chapter 5, some of the smaller particles may tend to be displaced from in front of the larger particle by fluid forces set up by the larger particle so that only a fraction of the particles contained in the volume will actually be collected (the fluid flow pattern is considered to be characterized by the flow around the large particle, taken to be a sphere). A knowledge of this fraction ϵ, known as either the efficiency or capture coefficient, thus permits solution of the problem since the number collected per unit time by a single particle $\mathcal{N}$, becomes

$$\mathcal{N} = \frac{\pi}{4} \epsilon d_L^2 v_L c \tag{16.24}$$

where c is the number concentration per unit volume of the smaller particles.

EXAMPLE 16.7

A 50-μm-diameter particle falls through an aerosol of 0.1-μm-diameter spheres containing 10^3 particles/cm^3. If the capture coefficient is 0.01, determine the number of particles collected in unit time.

From Table 6.1,

$$V_g = \tau g = (7.67 \times 10^{-3})(980) = 7.52 \text{cm/sec}$$

$$\mathcal{N} = \frac{\pi}{4}(0.01)(5 \times 10^{-3})^2(7.52)(10^3) = 1.48 \times 10^{-3} \text{particles/sec}$$

**ORDERED COAGULATION OF PARTICLES
OF APPROXIMATELY THE SAME SIZE**

For equal-sized particles prediction of a flow field is extremely difficult because the combined flows of both particles must be considered. These flow fields will vary, as pointed out earlier, as the particles approach each other, in addition to varying with Reynolds number and relative particle size.

An attempt at modeling this situation for viscous flow was made by Hocking (1960), who arrived at the conclusion that collisions between similarly-sized particles are impossible for particles smaller than 36 μm in diameter. Hocking's computations have been criticized on a number of grounds and it now appears that a small, but finite, impaction efficiency does exist. A rough approximation for ϵ in viscous flow has been given by Friedlander (1965),

$$\epsilon = \left(\frac{d_S}{d_L}\right)^2 - \left(\frac{d_S}{d_L}\right)^4 \tag{16.25}$$

SONIC AGGLOMERATION

Particle agglomeration can occur in ordered flow such as that which can be established in a sonic field. Here agglomeration takes place by the different velocities imparted to particles of differing inertia, by aerodynamic attractive forces between the particles and by radiation pressure which moves the particles towards the vibration antinodes. No complete and adequate theory for acoustic agglomeration as yet exists.

TURBULENT AGGLOMERATION

For turbulent agglomeration two cases should be considered. First, if the inertia of the aerosol particles is approximately the same as the medium, the particles will move about with the same velocities as associated air parcels and can be characterized by a turbulence or eddy diffusion coefficient D_T. This coefficient can have a value 10^4 to 10^6 times greater than aerosol diffusion coefficients. Turbulent agglomeration processes can be treated in a manner similar to conventional coagulation except that the larger diffusion coefficients are used.

The second method for aerosol coagulation in turbulent flows arises because of inertial differences between particles of different sizes. The particles accelerate to different velocities by the turbulence depending on their size, and may then collide with each other. This mechanism is unimportant for a monodisperse aerosol. For a polydisperse aerosol of unspecified size distribution Levich(1962) has shown that the agglomeration rate is proportional to the basic velocity of the turbulent flow raised to the 9/4 power, indicating that the agglomeration rate increases very rapidly with the turbulent velocity. Since very small particles are rapidly accelerated, this mechanism also decreases in importance as the particle size becomes very small, being most important for particles whose sizes exceed 10^{-5} to 10^{-4} cm in diameter. In all cases Brownian diffusion predominates when particles are less than 10^{-6} cm in diameter.

PROBLEMS

1. If coagulation is considered to be independent of particle size, how long will it take an aerosol concentration of 5×10^7 particles/cm^3 to coagulate to one half its original value? On the average, how many collisions per particle will have taken place?

2. Sulfur dioxide reacts rapidly in the atmosphere to form sulfuric acid droplets with an initial diameter of 0.001 μm. Determine a value for K_0 for coagulation of two of these particles.

3. Compute the value for the coagulation coefficient for 0.07-μm and 3.0-μm diameter particles.

4. Particles 1-μm in diameter are coagulating in an atmosphere where the temperature is $-85°C$ and the barometric pressure is 0.5 torr. Compute the value of the coagulation coefficient. Would you expect a greater or lesser rate of coagulation than at ambient conditions?

5. Estimate the time for the concentration of a test aerosol to be reduced by half if the aerosol concentration is 50 mg/m^3, its mean particle diameter is 0.1 μm and $\sigma_g = 2.1$.

REFERENCES

Abramowitz, M., and Stegun, I., *Handbook of Mathematical Functions, AMS 55*, National Bureau of Standards, Washington, D.C. (1964).

Aerosol Technology Committee, *A.I.H.A.J.*, **31**, 133 (1970).

Agricola, Georgius, *De Re Metallica*, translated by H. C. Hoover and L. H. Hoover, Dover Publications, New York (1950).

Aitken, J., *Proc. Royal Soc.*, **A51**, 408 (1892).

Aitken, J., *Trans. Royal Soc. Edinb.*, **30**, 337 (1880).

Amelin, A. G., *Theory of Fog Condensation*, Israel Program for Science Trans., Jerusalem (1967).

Barrer, R. M., *Diffusion In and Through Solids*, Cambridge University Press, Cambridge (1941).

Berry, J., *J. Opt. Soc. Amer.*, **52**, 888 (1962).

Berry, J., *J. Opt. Soc. Amer.*, **56**, 460 (1966).

Billings, C. E., et al., *Controlling Airborne Particles*, National Academy of Sciences, Washington, D.C. (1980).

Bird, R. B., Stewart, W. E., and Lightfoot, E. N., *Transport Phenomena*, Wiley, New York (1960).

Bodaszewski, R. W., *Beibl.*, **8**, 488 (1883).

Boisdron, Y., and Brock, J. R., *Atm. Env.*, **3**, 235 (1969).

Bricard, J., *Geofisica pura e appl.*, **51**, 237 (1962).

Bricard, J., and Pradel, J., in *Aerosol Science*, C. N. Davies, Ed., Academic, New York (1966).

Bricard, J., in *Problems of Atmospheric and Space Electricity*, S. C. Coroniti, Editor, Elsevier, Amsterdam (1965).

Brown, R., *Phil. Mag.*, **4**, 161 (1828).

Byers, H. R., *Elements of Cloud Physics,* University of Chicago Press, Chicago (1965a).

Byers, H. R., *Ind. Eng. Chem.,* **57**, 11 (1965b).

Cadle, R. D., *Particles in the Atmosphere and Space,* Reinhold, New York (1966).

Chan, T., and Lippmann, M., *Environ. Sci. Tech.,* **11**, 372 (1977).

Chapman, S., and Cowling, T. G., *The Mathematical Theory of Non-Uniform Gases,* Cambridge University Press, Cambridge (1961).

Charlson, R. J., Horvath, H., and Pueschel, R. F., *J. Atm. Env.,* **1**, 469 (1967).

Clark, W. E., and Whitby, K. T., *J. Atm. Sci.,* **26**, 603 (1967).

Connor, W. D., and Hodkinson, J. R., *Observations on the Optical Properties and Visual Effects of Smoke Plumes,* PHS Pub. 949-AP-30, Washington, D.C. (1967).

Cooper, D. G., and Reist, P. C., *J. Col. Int. Sci.,* **45**, 17 (1973).

Coulier, M., *J. Pharm. Chim.,* **22**, 165 (1875).

Crawford, M., *Air Pollution Control Theory,* McGraw-Hill, New York (1976).

Cunningham, E., *Proc. Roy. Soc. A,* **83**, 357 (1910).

Daniels F. and Alberty, R., *Physical Chemistry, 2nd Edition,* Wiley, New York (1961).

Davies, C. N., *Ann. Occ. Hyg.,* **8**, 239 (1965).

Davies, C. N., *Aerosol Sci.,* **10**, 477 (1979).

Davies, C. N., in *Aerosol Science,* C. N. Davies, Ed., Academic, New York (1966).

Davies, C. N., in *Fundamentals of Aerosol Science,* D. T. Shaw, Ed., Wiley, New York (1978).

Deirmendjian, D., *Electromagnetic Scattering on Spherical Polydispersions,* Elsevier, New York (1969).

Denman, H. H., Heller, W., and Pangonis, W. J., *Angular Scattering Functions for Spheres,* Wayne State University Press, Detroit (1966).

Dennis, R., *Handbook on Aerosols,* TID-26608, NTIS, Springfield, Va. (1976).

Deutsch, W., *Ann. d. Phys.,* **68**, 335 (1922).

Drinker, P. D., and Hatch, T., *Industrial Dusts, 2nd Edition,* McGraw-Hill, New York (1954).

Einbinder, H., *J. Chem. Phys.,* **26**, 948 (1957).

Einstein, A., *Investigations on the Theory of Brownian Movement,* Dover, New York (1956).

Exner, *Sitzungsber Akad. Wiss. Wien,* **56**, Part II, 116 (1867).

Flanagan, V. P. V., and O'Connor, T. C., *Geofis, Pura Appl.,* **50**, 148 (1961).

Fleagle, R. G., and Businger, J. A., *An Introduction to Atmospheric Physics,* Academic, New York (1963).

Fletcher, F. W., *J. Chem. Phys.,* **31**, 1136 (1958a).

Fletcher, F. W., *J. Chem. Phys.,* **29**, 572 (1958b).

Foitzik, I., Hebermehl, G., and Spankuch, D., *Optik,* **23**, 274 (1965/1966).

Fox, D. L., Kuhlman, M. R., and Reist, P. C., *Colloid and Interface Science,* Vol. II, M. Kerker, Ed., Academic, New York (1976).

Friedlander, S. K., and Wang, C. S., *J. Col. Int. Sci.,* **22**, 126 (1966).

Friedlander, S. K., in *Aerosols, Physical Chemistry and Applications,* C. Spurny, Ed., Gordon and Breach, New York (1965).

Fuchs, N. A., *Izv. Akad. Nauk. SSSR, Ser. Geogr. Geogfiz.,* **11**, 341 (1957).

Fuchs, N. A., *Geofis. Pura Appl.,* **56**, 185 (1963).

Fuchs, N. A., *The Mechanics of Aerosols,* Pergamon, New York (1964).

Fuchs, N. A., in *Fundamentals of Aerosol Science,* David T. Shaw, Ed., Wiley, New York (1978).

Fuchs, N. A., *Z. Phys.,* **89**, 736 (1934).

Fuchs, N. A., *Evaporation and Droplet Growth in Gaseous Media,* Pergamon, London (1959).

Giese, R., De Bary, E., Bullrich, K., and Vinnemann, C. V., Tabellen der Streufunktionen $i_1(\theta)$, $i_2(\theta)$, Akademie-Verlag, Berlin (1962).

Gillespie, T., *J. Col. Sci.,* **18**, 562 (1963).

Goetz, A. and Pueschel, R., *Atmos. Env.,* **1**, 287 (1967).

Gormley, P. G., and Kennedy, M., *Proc. Royal Irish Acad.*, **52A**, 163 (1949).

Green, H. L., and Lane, W. R., *Particulate Clouds: Dusts, Smokes and Mists*, 2nd ed., Van Nostrand, New York, (1964).

Gucker, F. T., and O'Konski, C. T., *J. Col. Sci.*, **4**, 541 (1949).

Happel, J., and Brenner, H., *Low Reynolds Number Hydrodynamics*, Prentice-Hall, Englewood Cliffs (1965).

Herdan, G., *Small Particle Statistics*, 2nd ed., Academic, New York (1960).

Hidy, G. M., *J. Col. Sci.*, **20**, 123 (1965).

Hidy, G. M., and Brock, J. R., *The Dynamics of Aerocollidal Systems*, Pergamon, Oxford (1970).

Hochrainer, D., Hidy, G. M., and Zebel, G., *J. Col. Sci.*, **30**, 553 (1969).

Hochrainer, D., and Zebel, G., *J. Aerosol Sci.*, **12**, 49 (1981).

Hocking, L. M., in *Aerodynamic Capture of Particles*, E. G. Richardson, Ed., Pergamon, Oxford (1960).

Hodkinson, J. R., in *Aerosol Science*, C. N. Davies, Ed., Academic, New York (1966).

Johnson, I., and LaMer, V. K., *J. Am. Chem. Soc.*, **69**, 1184 (1947).

Joos, G., *Theoretical Physics*, 2nd ed., Blackie and Son, Glasgow (1951).

Jost, W., *Diffusion in Solids, Liquids and Gases*, Academic, New York (1952).

Junge, C., *Air Chemistry and Radioactivity*, Academic, New York (1963).

Junge, C., *Tellus*, **5**, 1 (1953).

Keefe, D., Nolan, P. J., and Rich, T. A., *Proc. Royal Irish Acad.*, **60A**, 27 (1959).

Kerker, M., *The Scattering of Light and Other Electromagnetic Radiation*, Academic, New York (1969).

Kerker, M., *Electromagnetic Scattering*, Pergamon, New York (1963).

Kitani, S., *J. Col. Sci.*, **15**, 287 (1960).

Koch, W., *J. Aerosol Sci.*, **13**, 415 (1982).

Koschmieder, H., *Beitr. Phys. freien Atm.*, **12**, 33 (1924a).

Koschmieder, H., *Beitr. Phys. freien Atm.*, **12**, 171 (1924b).

Kuhlman, M. R., *A Model of Sulfuric Acid Aerosol Processes*, Ph.D. dissertation, University of North Carolina, Chapel Hill (1982).

Landau, L. D. and Lifshitz, E. M., *Fluid Mechanics*, Pergamon, New York (1959).

Lanza, A. J., in *Silicosis and Asbestosis*, Oxford University Press, New York (1938).

Lawton, J., and Weinberg, F. J., *Electrical Aspects of Combustion*, Clarendon Press, Oxford (1969).

Leith, D., and Licht, W., *A.I.Ch.E. Symp. Ser.*, **126**, 196 (1972).

Lenard, P., *Ann. der Phys.*, **47**, 413 (1915).

Levich, V. G., *Physicochemical Hydrodynamics*, Prentice-Hall, Englewood Cliffs (1962).

Levin, L. M., *Physics of Coarse Aerosols*, Trans. 815, U.S. Dept. of Commerce TT 64 19550 (June 1963).

Licht, W., *Air Pollution Control Engineering*, Marcel Dekker, New York (1980).

Lippman, M., *A.I.H.A.J.*, **31**, 138 (1970).

Liu, B. H. Y., and Pui, D. Y. H., *J. Col. Int. Sci.*, **58**, 142 (1977).

Liu, B. Y. H., Whitby, K. T., and Yu, H. S., *J. Col. Int. Sci.*, **23**, 367 (1967a).

Liu, B. Y. H., Whitby, K. T., and Yu, H. S., *J. Appl. Phys.*, **38**, 1592 (1967b).

Lowan, A. N., *Tables of Scattering Functions for Spherical Particles*, NBS Ap. Math Ser. **4**, Washington, D.C. (1948).

Maron, S. H., and Elder, M. E., *J. Col. Sci.*, **18**, 199 (1963).

Marple, V. A., *A Fundamental Study of Inertial Impactors*, Ph.D. dissertation, University of Minnesota, Part. Tech. Lab. Pub. No. 144 (1970).

Marple, V. A., and Willeke, K., in *Aerosol Measurement*, D. A. Lundgren et al. Eds., University of Florida Press, Gainesville (1979).

Mason, B. J., *The Physics of Clouds*, 2nd ed., Clarendon Press, Oxford (1971).

Maxwell, J. C., *Encyclopedia Britannica, Xth Edition*, **xx**, xx, (1877).

McDaniel, E. W., *Collision Theory in Ionized Gases*, Wiley, New York (1964).

Mercer, T. T., *Aerosol Technology and Hazard Evaluation*, Academic, New York (1973).

Middleton, W. E. K., *Vision Through the Atmosphere*, University of Toronto Press, Toronto (1963).

Mie, G., *Ann. Physik.*, **25**, 377 (1908).

Miller, F. C., and Loeb, L. B., *J. Appl. Phys.*, **22**, 614 (1951b).

Miller, F. C., and Loeb, L. B., *J. Appl. Phys.*, **22**, 740 (1951c).

Miller, F. C., and Loeb, L. B., *J. Appl. Phys.*, **22**, 494 (1951a).

Miller, J. G., and Heinemann, H., *Science*, **107**, 144 (1948).

Millikan, R. A., *Phys. Rev, 2nd Ser.*, **22**, 1 (1923).

Millikan, R. A., *Science*, **32**, 436 (1910).

Mokler, B., *personal communication* (1969).

Muller, H., *Kolliodchem. Beih.*, **27**, 223 (1928).

Napper, D. H., and Ottewill, R. H., *Trans. Faraday Soc.*, **60**, 1466 (1964).

Natanson, G. L., *Soviet Phys., tech. Phys.*, **5**, 538 (1960).

Nolan, P. J., and Doherty, D. J., *Proc. Royal Irish Acad.*, **53A**, 163 (1950).

O'Connor, T. C., and Sharkey, W. P., *Proc. Royal Irish Acad.*, **61A**, 15 (1960).

Orr, C., Hurd, F. K., and Corbett, W. J., *J. Col. Sci.*, **13**, 472 (1958).

Orr, C., et al., *J. Met.*, **15**, 240 (1958).

Orr, C., *Particulate Technology*, Macmillan, New York (1966).

Pauthenier, M. M. and Moreau-Hanot, M., *J. Phys. Rad., Ser. 7*, **3**, 590 (1932).

Pendorf, R., *J. Opt. Soc. Am.*, **52**, 797 (1962).

Perry, R. H., and Chilton, C. H., *Chemical Engineers Handbook, 5th Ed.*, McGraw-Hill, New York (1973).

Pradtl, L., and Tietjens, O. G., *Fundamentals of Hydro- and Aeromechanics*, Dover, New York (1957).

Pruppacher, H. R., and Klett, J. D., *Microphysics of Clouds and Precipitation*, Reidel, Boston (1978).

Raabe, O. G., *J. Aerosol Sci.*, **2**, 289 (1971).

Ranz, W. E., and Wong, J. B., *Industr. Eng. Chem.*, **44**, 1371 (1952).

Rao, A. K., and Whitby, K. T., *J. Aerosol Sci.*, **9**, 77 (1978).

Rayleigh, Lord, *Proc. Roy. Soc.*, **34**, 414 (1882).

Rayleigh, Lord, *Phil. Mag.*, **47**, 375 (1899).

Robinson, R. A., and Stokes, R. H., *Electrolyte Solutions*, Butterworth, London (1959).

Rohmann, H., *Zeit. Phys.*, **17**, 253 (1923).

Rose, H. E., and Wood, A. J., *An Introduction to Electrostatic Precipitation in Theory and Practice*, Constable and Company, London (1966).

Sartor, J. D., and Atkinson, W. R., *Science*, **157**, 1267 (1967).

Schlichting, H., *Boundary Layer Theory, 6th Ed.*, McGraw Hill, New York (1968).

Seddig, R., *Phys. Zeit.*, **9**, 465 (1908).

Sehmel, G. A., *Aerosol Deposition from Turbulent Airstreams in Vertical Conduits*, BNWL-578, Richland, Wash. (1968).

Sinclair, D., *A.I.H.A.J.*, **33**, 729 (1972).

Sinclair, D., in *Handbook on Aerosols*, H. T. Johnstone, Ed., U.S. Govt. Printing Office, Washington, D.C. (1950).

Sliney, D. and Wolbarsht, M., *Safety With Lasers and Other Optical Sources*, Plenum, New York (1980).

Smoluchowski, M., *Bull. Acad. Sci., Cracow*, **1a**, 28 (1911).

Smoluchowski, M., *Zeit. Phys. Chemie*, **92**, 129 (1918).

Smoluchowski, M., *Zeit. Phys. Chemie*, **92**, 167 (1918).

Spurny, K. R., Gentry, J. W., and Stober, W., in *Fundamentals of Aerosol Science*, D. T. Shaw, Ed., Wiley, New York (1978).

Stober, W., and Flachsbart, H., *Env. Sci. Tech.,* **3,** 1280 (1969).

Stober, W., in *Assessment of Airborne Particles,* T. T. Mercer, T. E. Morrow, and W. Stober, Eds., Charles Thomas, Philadelphia (1972).

Sutton, O. G., *Mathematics in Action,* 2nd ed., G. Bell and Sons, London (1957).

Sutugin, A. G., *Russian Chem. Rev.,* **38,** 79 (1969).

Svedberg, Z. *Electroch.,* **12,** 853 (1909).

Takahashi, K., and Kasahara, A., *Atm. Env.,* **2,** 441 (1968).

Tikhomirov, M., Tunitskii, N. and Petryanov, I., *Dokl. Akad. nauk SSSR,* **94,** 865 (1942).

Tillery, M. I., *Aerosol Measurement,* Dale Lundgren et al. Ed., University Presses of Florida, Gainesville (1979).

Tricker, R. A., *Introduction to Meteorologcal Optics,* Elsevier, New York (1970).

Van de Hulst, H. C., *Light Scattering by Small Particles,* Wiley, New York (1957).

Wait, G. R., *Carnegie Inst. of Wash. News Serv. Bul.,* **3,** 1 (1934).

Wang, C. S. and Friedlander, S. K., *J. Col. Int. Sci.,* **24,** 170 (1967).

Weast, R. C., Ed., *Handbook of Chemistry and Physics, 54th Ed.,* CRC Press, Cleveland (1973).

Went, F. W., *Proc. Nat. Acad. Sci.,* **46,** 212 (1960).

Whitby, K. T., and Peterson, C. M., *Ind. Eng. Chem. Fund.,* **4,** 66 (1965).

Whitby, K. T., and Willeke, K., in *Aerosol Measurement,* D. A. Lundgren, et al., Eds., University of Florida Press, Gainesville, (1979).

Whitby, K. T., and Liu, B. Y. H., *J. Col. Int. Sci.,* **25,** 537 (1967).

Whitby, K. T., and Liu, B. Y. H., in *Aerosol Science,* C. N. Davies, Ed., Academic, New York (1966).

Whitby, K. T., *Rev. Sci. Inst.,* **32,** 1351 (1961).

White, H. J., J., *APCA,* **27,** 206 (1977).

White, H. J., *Industrial Electrostatic Precipitation,* Addison-Wesley, New York (1963).

Whytlaw-Gray, R., and Patterson, H. S., *Smoke,* E. Arnold, London (1932).

Wilson, C. T. R., *Phil. Trans. Royal Soc.,* **A189,** 265 (1897).

Yaglou, C. P. and Benjamin, L. C., *Heating, Piping and Air Conditioning,* Jan., 25 (1934).

Zak, E. G., *Zh. Geofiz.,* **6,** 452 (1936).

Zebel, G., in *Aerosol Science,* C. N. Davies, Ed., Academic, New York (1966).

CORRECTED SEDIMENTATION VELOCITIES

**Unit density spheres
(NTP)**

d (μm)	v_g (cm/sec)	C_c
0.01	7.07×10^{-6}	23.775
0.02	1.45×10^{-5}	12.192
0.04	3.06×10^{-5}	6.422
0.06	4.84×10^{-5}	4.516
0.08	6.80×10^{-5}	3.573
0.10	8.97×10^{-5}	3.015
0.20	2.31×10^{-4}	1.938
0.40	6.88×10^{-4}	1.446
0.60	1.39×10^{-3}	1.294
0.80	2.32×10^{-3}	1.220
1.00	3.50×10^{-3}	1.176
2.00	1.29×10^{-2}	1.088
4.00	4.97×10^{-2}	1.044
6.00	1.10×10^{-1}	1.029
8.00	1.94×10^{-1}	1.022
10.00	3.03×10^{-1}	1.018
20.00	1.20×10^{0}	1.009
40.00	4.78×10^{0}	1.004
60.00	1.07×10^{1}	1.003

STOKES' LAW

Stokes considered the resistance experienced by a sphere moving uniformly through an incompressible viscous fluid. Viscous flow implies a low Re. He assumed an infinite medium, rigid particle, no slipping at surface of particle. Stokes pointed out that his solution was erroneous in the case of a cylinder.

Consider a fluid flowing past a sphere. Take spherical coordinates with the polar axis parallel to u, the direction of motion of the fluid. By symmetry, all quantities of interest will only be functions of r and the polar angle θ.

The resisting force $\vec{F}$ is parallel to the velocity $\vec{u}$. The incremental force acting on a unit of surface area is

$$d\vec{F}_i = p\vec{n}_i - \sigma_{ik}\vec{n}_k$$

where p is the pressure of the fluid, σ_{ik} is the viscosity stress tensor, and n is a unit vector. The term σ describes the irreversible "viscous" transfer of momentum in the field. Hence the first term in the above equation is the ordinary pressure of the fluid, the second represents the force of friction, due to viscosity, acting on the surface of the body.

Considering only the components of these forces in the $\vec{u}$ direction, and summing over the entire sphere surface gives

$$F = \oint (-p \cos \theta + \sigma_{rr} \cos \theta - \sigma_{r\theta} \sin \theta)ds$$

where the integration is taken over the whole surface, s, of the sphere. We must now evaluate p, σ_{rr} and $\sigma_{r\theta}$. In spherical coordinates the stress tensor is

$$\sigma_{rr} = 2u\,\frac{\partial v_r}{\partial r}$$

$$\sigma_{re} = \mu\left[\frac{1}{r}\frac{\partial v_r}{\partial \theta} + \frac{\partial v_\theta}{\partial r} - \frac{v_\theta}{r}\right]$$

where v is the velocity of a fluid element ($v = u$ at infinity) and μ is the fluid viscosity. For steady flow of an incompressible fluid the Navier-Stokes equation is

$$\vec{v} \times \operatorname{grad} \vec{v} = -\left(\frac{1}{\rho}\right)\operatorname{grad} p + \frac{u}{\rho}\nabla\vec{v}$$

the term

$$\vec{v} \times \operatorname{grad} \vec{v} \text{ is proportional to} \frac{u^2}{l}$$

and

$$\frac{\mu}{\rho}\nabla\vec{v} \text{ is proportional to } \frac{\mu u}{\rho l^2}$$

The ratio of the two is the Reynolds' number, Re, so if Re is small we can write

$$\mu\,\nabla\vec{v} - \operatorname{grad} p = 0$$

which with the equation of continuity

$$\operatorname{div} v = 0$$

completely specifies the motion. Solving these equations gives

$$v_r = u\,\cos\theta\left[1 - \frac{3R}{2r} + \frac{R^3}{2r^3}\right]$$

$$v_\theta = -u\,\sin\theta\left[1 - \frac{3R}{4r} - \frac{R^3}{4r^3}\right]$$

where the surface of the sphere is $r = R$.
The relative pressure is

$$p = -\left(\frac{3u}{2R}\right)u\,\cos\theta$$

substituting, at $r = R$

$$\sigma_{rr} = 0,\ \sigma_{r\theta} = -\left(\frac{3\mu}{2R}\right)u\,\sin\theta$$

thus

$$F = \frac{3\mu u}{2R} \oint ds$$

or

$$F = 6\pi R\mu u = 3\pi\mu u d$$

where d is the particle diameter.

For a more complete derivation the reader is referred to Landau and Lifshitz (1969) or Joos (1951).

ERROR FUNCTION

Many integrals of the diffusion equation lead to the error function which can be written as:

$$\text{erf } y = \frac{2}{\sqrt{\pi}} \int_0^y \exp(-\xi^2)\, d\xi$$

$$\text{erf}(\infty) = 1$$

$$\text{erf}(0) = 0$$

$$\text{erf}(-y) = -\text{erf}(y)$$

The first two derivatives of the error function are

$$\frac{d}{dy}(\text{erf } y) = \frac{2}{\sqrt{\pi}} \exp(-y^2)$$

$$\frac{d^2}{d^2 y}(\text{erf } y) = -\frac{4}{\sqrt{\pi}} y \exp(-y^2)$$

Table C-1 lists erf(y) for various values of the argument y. In this table y to the nearest tenth is entered in the left-hand column reading across in hundredths to the argument value of erf(y). Thus erf(0.63) = 0.62705.

Table C-1. Error function, erf (y)

y	0.00	0.01	0.02	0.03	0.04	0.05	0.06	0.07	0.08	0.09
0.0	0.00000	0.01128	0.02256	0.03384	0.04511	0.05637	0.06762	0.07886	0.09008	0.10128
0.1	0.11246	0.12362	0.13476	0.14587	0.15695	0.16800	0.17901	0.18999	0.20094	0.21184
0.2	0.22270	0.23352	0.24430	0.25502	0.26570	0.27633	0.28690	0.29742	0.30788	0.31828
0.3	0.32863	0.33891	0.34913	0.35928	0.36936	0.37938	0.38933	0.39921	0.40901	0.41874
0.4	0.42839	0.43797	0.44747	0.45689	0.46623	0.47548	0.48466	0.49375	0.50275	0.51167
0.5	0.52050	0.52924	0.53790	0.54646	0.55494	0.56332	0.57162	0.57982	0.58792	0.59594
0.6	0.60386	0.61168	0.61941	0.62705	0.63459	0.64203	0.64938	0.65663	0.66378	0.67084
0.7	0.67780	0.68467	0.69143	0.69810	0.70468	0.71116	0.71754	0.72382	0.73001	0.73610
0.8	0.74210	0.74800	0.75381	0.75952	0.76514	0.77067	0.77610	0.78144	0.78669	0.79184
0.9	0.79691	0.80188	0.80677	0.81156	0.81627	0.82089	0.82542	0.82987	0.83423	0.83851
1.0	0.84270	0.84681	0.85084	0.85478	0.85865	0.86244	0.86614	0.86977	0.87333	0.87680
1.1	0.88021	0.88353	0.88679	0.88997	0.89308	0.89612	0.89910	0.90200	0.90484	0.90761
1.2	0.91031	0.91296	0.91553	0.91805	0.92051	0.92290	0.92524	0.92751	0.92973	0.93190
1.3	0.93401	0.93606	0.93807	0.94002	0.94191	0.94376	0.94556	0.94731	0.94902	0.95067
1.4	0.95229	0.95385	0.95538	0.95686	0.95830	0.95970	0.96105	0.96237	0.96365	0.96490
1.5	0.96611	0.96728	0.96841	0.96952	0.97059	0.97162	0.97263	0.97360	0.97455	0.97546
1.6	0.97635	0.97721	0.97804	0.97884	0.97962	0.98038	0.98110	0.98181	0.98249	0.98315
1.7	0.98379	0.98441	0.98500	0.98558	0.98613	0.98667	0.98719	0.98769	0.98817	0.98864
1.8	0.98909	0.98952	0.98994	0.99035	0.99074	0.99111	0.99147	0.99182	0.99216	0.99248
1.9	0.99279	0.99309	0.99338	0.99366	0.99392	0.99418	0.99443	0.99466	0.99489	0.99511
2.0	0.99532	0.99552	0.99572	0.99591	0.99609	0.99626	0.99642	0.99658	0.99673	0.99688
2.1	0.99702	0.99715	0.99728	0.99741	0.99753	0.99764	0.99775	0.99785	0.99795	0.99805
2.2	0.99814	0.99822	0.99831	0.99839	0.99846	0.99854	0.99861	0.99867	0.99874	0.99880
2.3	0.99886	0.99891	0.99897	0.99902	0.99906	0.99911	0.99915	0.99920	0.99924	0.99928
2.4	0.99931	0.99935	0.99938	0.99941	0.99944	0.99947	0.99950	0.99952	0.99955	0.99957
2.5	0.99959	0.99961	0.99963	0.99965	0.99967	0.99969	0.99971	0.99972	0.99974	0.99975
2.6	0.99976	0.99978	0.99979	0.99980	0.99981	0.99982	0.99983	0.99984	0.99985	0.99986
2.7	0.99987	0.99987	0.99988	0.99989	0.99989	0.99990	0.99991	0.99991	0.99992	0.99992
2.8	0.99992	0.99993	0.99993	0.99994	0.99994	0.99994	0.99995	0.99995	0.99995	0.99996
2.9	0.99996	0.99996	0.99996	0.99997	0.99997	0.99997	0.99997	0.99997	0.99997	0.99998

ELECTRICAL UNITS

Coulomb's Law can be written as:

$$F = \frac{\gamma q q'}{4\pi r^2}$$

For cgs system

$$\gamma = \frac{4\pi}{\epsilon}$$

Then F is in dynes when q is expressed in statcoulombs (esu) and r is in centimeters.

To convert from cgs to mks units, start with the basic relationship

$$\frac{\text{cgs unit charge (in esu units)}}{\text{mks unit charge (in coulombs)}} = \frac{c}{10}$$

where c is the velocity of light = 2.99793×10^{10} cm/sec.

Then

$$4.8 \times 10^{-10} \text{ esu} = 1.6 \times 10^{-19} \text{ coulombs}$$

For mks system

$$\gamma = \frac{1}{k_0 \epsilon}$$

when F is expressed in Newtons, q in coulombs and r in meters,

$$k_0 = 8.8542 \times 10^{-12}$$

To convert to other units:

$$E = \frac{F}{q}, E(x) = \frac{(F/10^5)}{(10q/c)}, x = \frac{c}{10^6}$$

that is, multiply static field strength by $\approx 30,000$ to get mks field strength.

$$V = E \times l, \ V(x) = E \times 30,000 \times \frac{1}{100}, x = 300$$

that is, multiply cgs voltage by 300 to get mks voltage.

Table D-1. Conversion factors for various electrical parameters

Electrical unit	Absolute unit (mks)	Electrostatic unit (cgs)
Charge $(e^{1/2}m^{1/2}l^{3/2}t^{-1})$	1 coulomb	3×10^9 statcoulombs
Current $(e^{1/2}m^{1/2}l^{3/2}t^{-2})$	1 ampere	3×10^9 statamperes
Potential $(e^{-1/2}m^{1/2}l^{1/2}t^{-1})$	1 volt	3.34×10^{-3} statvolts
Resistance $(e^{-1}l^{-1}t)$	1 ohm	1.11×10^{-12} statohms

ADIABATIC EXPANSION

The vapor pressure of a substance at temperature T can be estimated from the expression given by Daniels and Alberty (1963) where the factors A and B are constants that depend on the substance under consideration:

$$\log P = A - \frac{B}{T} \qquad (E.1)$$

Now letting
P_1 = vapor pressure before expansion
P_2 = partial pressure before expansion
P_3 = partial pressure after expansion
P_4 = vapor pressure after expansion
Then S, the saturation ratio, is

$$S = \frac{P_3}{P_4} \qquad (E.2)$$

Furthermore, the temperature of a volume of gas after expansion can be estimated from

$$T_A = \frac{T_B}{R^{\gamma - 1}} \qquad (E.3)$$

where A = after, B = before, R is the expansion ratio defined as

$$R = \text{expansion ratio} = \frac{\text{final volume}}{\text{initial volume}} = \frac{V_A}{V_B} \qquad (E.4)$$

and γ is

$$\gamma = \frac{\text{critical pressure specific heat}}{\text{critical volume specific heat}}$$

Since $P_1 V_1/T_1 = P_2 V_2/T_2$,

$$P_3 = P_2 \frac{1}{R} \frac{T_A}{T_B} \qquad (E.5)$$

If it is assumed that $P_2 = P_1$, that is saturated conditions exist before expansion, then

$$S = \left(\frac{V_2}{V_1}\right) \exp\left[\frac{\Phi}{T_B}\left(\frac{V_2}{V_1}\right)^{\gamma-1} - 1\right] \qquad (E.6)$$

For water, $\gamma = 1.4$ and $\Phi = 5343$, give the following:

Expansion ratio	Saturation ratio
1.0	1.00
1.1	1.78
1.2	3.08
1.3	5.21
1.4	8.64

PSYCHROMETRIC CHART

PSYCHROMETRIC CHART

Normal Temperatures

Reproduced Courtesy of Carrier Corporation

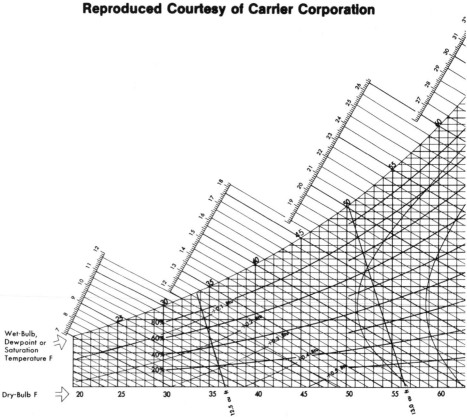

Wet-Bulb, Dewpoint or Saturation Temperature F

Dry-Bulb F

Below 32 F, properties and enthalpy deviation lines are for ice.

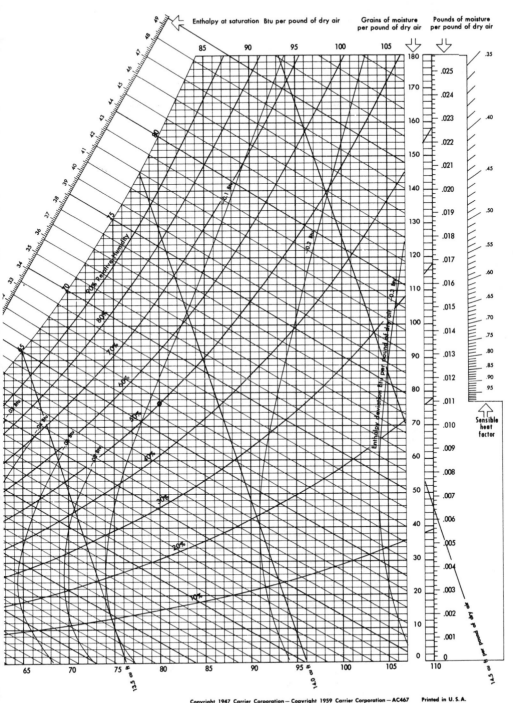

Enthalpy at saturation Btu per pound of dry air

Grains of moisture per pound of dry air

Pounds of moisture per pound of dry air

Sensible heat factor

BESSEL FUNCTIONS OF ORDER 1[a]

x	$J_1(x)$	x	$J_1(x)$
0.0	0.00000	2.0	0.57672
0.1	0.04994	2.1	0.56829
0.2	0.09950	2.2	0.55596
0.3	0.14832	2.3	0.53987
0.4	0.19603	2.4	0.52019
0.5	0.24227	2.5	0.49709
0.6	0.28670	2.6	0.47082
0.7	0.32900	2.7	0.44160
0.8	0.36844	2.8	0.40971
0.9	0.40595	2.9	0.37543
1.0	0.44005	3.0	0.33906
1.1	0.47090	3.1	0.30092
1.2	0.49829	3.2	0.26134
1.3	0.52202	3.3	0.22066
1.4	0.54195	3.4	0.17923
1.5	0.55794	3.5	0.13738
1.6	0.56990	3.6	0.09547
1.7	0.57777	3.7	0.05383
1.8	0.58152	3.8	0.01282
1.9	0.58116		

[a]From M. Abramowitz and I. Stegun, Handbook of Mathematical Functions, Nat. Bureau of Standards Appl. Math Ser. 55, Washington, D.C., 1964, p. 390.